UNIVERSITY OF NOTTINGHAM
10 0413197 5
WITHDRAWN
FROM THE LIBRARY

D1338632

THE TRANSMISSION-LINE MODELING METHOD

TLM

IEEE/OUP Series on Electromagnetic Wave Theory

The IEEE/OUP Series on Electromagnetic Wave Theory consists of new titles as well as reprintings and revisions of recognized classics that maintain long-term archival significance in electromagnetic waves and applications.

Books in the Series

Chew, W. C., *Waves and Fields in Inhomogeneous Media,* 1995

Christopoulos, C., *The Transmission-Line Modeling Method: TLM,* 1995

Collin, R. E., *Field Theory of Guided Waves,* Second Edition, 1991

Dudley, D. G., *Mathematical Foundations for Electromagnetic Theory,* 1994

Elliott, R. S., *Electromagnetics: History, Theory, and Applications,* 1993

Felsen, L. B. and Marcuvitz, N., *Radiation and Scattering of Waves,* 1994

Harrington, R. F., *Field Computation by Moment Methods,* 1993

Jones, D.S., *Methods in Electromagnetic Wave Propagation,* Second Edition, 1994

Lindell, I.V., *Methods for Electromagnetic Field Analysis,* 1992

Tai, C. T., *Generalized Vector and Dyadic Analysis: Applied Mathematics in Field Theory,* 1991

Tai, C. T., *Dyadic Green Functions in Electromagnetic Theory,* Second Edition, 1994

Van Bladel, J., *Singular Electromagnetic Fields and Sources,* 1991

THE TRANSMISSION-LINE MODELING METHOD

TLM

CHRISTOS CHRISTOPOULOS

UNIVERSITY OF NOTTINGHAM

IEEE Microwave Theory and Techniques Society, *Sponsor*

The Institute of Electrical
and Electronics Engineers, Inc.
New York

Oxford University Press
Oxford

IEEE PRESS
445 Hoes Lane, PO Box 1331
Piscataway, NJ 08855-1331

A NOTE TO THE READER
This book has been electronically reproduced from digital information stored at John Wiley & Sons, Inc. We are pleased that the use of this new technology will enable us to keep works of enduring scholarly value in print as long as there is a reasonable demand for them. The content of this book is identical to previous printings.

Library of Congress Cataloging-in-Publication Data

Christopoulos, Christos.
Transmission-line modeling method: TLM / by C. Christopoulos.
p. cm.—(IEEE/OUP series on electromagnetic wave theory)
Includes bibliographical references and index.
ISBN 0-7803-1017-9
1. Microwave transmission lines—Mathematical models. I. Title. II. Series.
TK7876.C57 1995
621.3–dc20 94-45582
CIP

To my parents
Kosta and Yiota
for supporting me for many years

Contents

Preface

This book is an attempt to present, in a systematic way, the development and application of the transmission-line modeling method. The method is commonly referred to as TLM and traces its rapid development from the early 1970s.

It is always difficult to judge the correct time to write a book on a rapidly changing topic. One is painfully aware of the advances made even while writing, and the impossibility of doing justice to all the changing strands of the subject. Yet, it is important to provide interested readers with a comprehensive view of TLM to help them make sense of the more than 300 publications related to this method, and to afford newcomers to the subject a rapid grasp of the fundamentals that will serve as an introduction to more advanced topics.

The development of TLM should be viewed as part of a general trend in recent years to supplement traditional experimental and analytical techniques with numerical simulation. The latter approach is dependent on advances in computer technology and the capacity of researchers to find new ways of constructing models for solution by digital computer. TLM is a method based on this approach, and it has many computational and conceptual advantages.

This book aims to be comprehensive, starting from basic transmission line theory and working through TLM discrete models of lumped components, including one-, two-, and three-dimensional problems. The emphasis is on electromagnetics, but other applications such as in thermal and acoustic problems are also considered. Chapter 9 contains a survey of application areas to guide the reader who may have an interest in a particular topic, while Chapter 10 is focused on more advanced topics in TLM.

The first five chapters should be accessible to a large number of readers, including advanced undergraduates. Certain topics may be omitted on first reading (i.e., Sections 2.3, 3.4 through 3.6, 4.3, and 5.3) without loss of continuity, and the remaining material can be presented in 20 lectures if required. Basic three-dimensional models can be introduced with a minimum of effort (Sections 6.1 and 6.2), but more general and flexible three-dimensional work requires mastery of the material in the remainder of Chapter 6, and also that of Chapter 10. Modifications to TLM to deal with problems in areas other than electromagnetics are described in Chapters 7 and 8. It is hoped that researchers and practitioners in simulation will find the book useful and will make their own contributions to this important engineering simulation method.

C. Christopoulos
Nottingham, England

Acknowledgments

It is a pleasure to acknowledge the contributions made to my understanding of this subject by a number of colleagues both in my immediate working environment and in research groups working around the world. More specifically, I am indebted to the late Prof. P. B. Johns for introducing me to TLM, and to my co-workers over the years, including Dr. P. Naylor, Dr. S. Y. R. Hui, Dr. J. L. Herring, Dr. D. D.Ward, Dr. M. da F. Mattos, Mr. R. Scaramuzza, Dr. A. P. Duffy, Mr. V. Trenkic, Dr. T. M. Benson, and Mr. D. P. Johns. These individuals provided numerous thought-provoking discussions and much practical help. I also wish to thank Miss S. E. Hollingsworth for competently typing large parts of the draft and making sense of my notation.

1

Introduction to Numerical Modeling

1.1 MODELING AS AN INTELLECTUAL ACTIVITY

Humans often attempt to understand physical phenomena by reduction to the familiar. At the end of the nineteenth century, scientists used models of electric and magnetic phenomena that were essentially mechanical in nature [1]. Mechanical phenomena at that time were accorded the status of familiar concepts and, by analogy to them, electric and magnetic phenomena were made plausible. Irrespective of their explanatory significance, it cannot be denied that models are very useful in several respects [2]. If the laws of a new phenomenon have the same form with those of another which has already been studied, then the consequences of the latter can be transferred to the new phenomena. This offers intellectual economy, strengthens the generality, and broadens the scope of our understanding of the world. A well chosen model facilitates the grasp of a new phenomenon and can be an effective heuristic tool in the search for explanations.

Numerical modeling is an activity distinct from computation. The endless printouts of numbers generated by computers are meaningless outside a system of knowledge and rationality that stems from human activity and experience. Numbers by themselves do not convey substantial information unless they can be used as evidence to justify or reject conceptual models of the phenomena we observe [3]. By repeated observations, we establish regularities which are embodied in models (mathematical or otherwise). This is a never-ending process which generates models of increasing generality and power. It is worth pointing out that however general and sophisticated a model is, it is never the real thing. Whether a perfect model is possible is a philosophical question of great complexity. However, a pragmatic answer to this question can be given which may be acceptable for engineering models. A model that is "perfect" in the sense of being identical to the real thing will be of limited use to the engineer and therefore undesirable. It will be as unwieldy as the real thing and will obscure the insights that a simple but effective model can often provide. The development of engineering models must therefore be a task focused on simplicity and clarity. Models must be simple—but not simpler than they have to be. It therefore follows that a task equally important as the development of a model is the identification, in a qualitative and quantitative manner, of its limitations. A consequence of this approach is that several different models are necessary to describe what is the same thing. An example of this can be seen in the modeling of a human being [4]. In embarking on this task the following question must be asked: "What is the intended application of the model?" If the answer to this question is that it is intended for clothes fitting, then a simple wooden structure padded with straw is perfectly adequate. If, on the other hand, the model is to be used for studying the effect on humans of rapid acceleration and vibrations, then a structure made of springs and masses will be the most suitable model. Similarly, a model for determining the body's electrostatic field distribution will need to contain a large amount of water to account for the basic constituent of the human body which, by virtue of its high dielectric permittivity, will affect strongly the field distribution. A model of humans that consists of resistors and capacitors is quite effective in predicting electrostatic charging and discharging. There is no way of saying that one of these models is a better model of a human, except in the sense that it is or may be developed into a more general model.

The above example helps to illustrate the great variety of models that can be constructed. We all have a multitude of mental models of which we may not be conscious at all times. It is simply impossible to have a rational view of the world without some models.

The reader may be persuaded by now of the importance of modeling in all human activities. However, modeling per se, as a field of study, is in its infancy, and there is simply a very wide scope for study and development of the modeling process.

Focusing now attention on numerical models, it is worth pointing out that implicit to every modeling activity are the following more or less distinct steps adapted from Reference [5].

a) *Conceptualization.* This is the first step in the modeling mental process whereby observations are related to relevant physical principles. (For example, I release an apple and it falls. This must have something to do with gravity and not with its color!)

b) *Formulation.* This step consists of the more detailed formulation of the physical ideas, perhaps in a mathematical form. (For example, in the example above, state Newton's Law **F** = m**a** and quantify other factors that may be thought to be relevant, such as air drag, etc.)

c) *Numerical implementation.* During this stage, the mathematical or other model described above is prepared for solution—most probably by a digital computer. A solution algorithm is developed that is suitable for implementation by the computer. This process can be a simple one (as for the apple example) or very complex (as with most problems in electromagnetics).

d) *Computation.* This stage involves the coding of the solution algorithm using one of the computer languages and the development of preprocessing and postprocessing facilities. Large programs may involve extensive number crunching and press computing facilities to their limit. Issues of computational efficiency must be considered carefully at this stage.

e) *Validation.* Modeling complex problems involves a number of simplifications and approximations. At any stage during the process described above, unacceptable errors may be introduced. It is not uncommon for users of powerful models and computers to regard their output results as beyond reproach. It is, however, essential that results be checked for physical reasonableness. Confidence in models must always be tempered by an understanding of the complexity of real problems.

This brief introduction to modeling should convince the reader of the skill and sophistication necessary for good modeling. Conceptualizing demands the skills of a physicist and engineer to identify relevant mechanisms and thus develop the necessary framework for a solution. The more detailed formulation of the problem requires mathematical skills, as does the numerical analysis necessary for algorithm development and implementation. Computation can always benefit from the contributions of a computer scientist and engineer. Finally, the special gifts of an experimentalist are required to do critical, well documented experiments with the required accuracy and to interpret results for comparison with simulations.

Naturally, no single person can claim mastery of all these fields. Nevertheless, anyone seriously involved in modeling should aim to develop the skills, maturity, and confidence necessary to function creatively and efficiently in this exciting discipline.

In this text, which is focused on a single modeling method, it is not possible to present in detail all aspects of modeling. However, by a systematic approach to model building in *transmission line modeling* (TLM), the basic processes inherent in many similar methods will be presented, and it is hoped that they will be of value in other applications and disciplines. The plan of the book is outlined below.

A general classification of modeling methods is given in the next section, based on the manner in which the problem is formulated. The chapter concludes with a section that aims to put TLM into context rather than offering a complete coverage of other methods.

The basic building blocks in TLM are electric circuit components and, more specifically, transmission line segments. The next two chapters give an introduction to the basic modeling philosophy and to standard transmission line theory. Subsequent chapters introduce model building from the simplest lumped components to the most general field distributions in three dimensions. The standard theory is presented in terms of application to electromagnetics. This avoids unnecessary complexity and loss of focus during the development of the basic concepts and techniques. Generalization to other applications (e.g., thermal, mechanical, etc.) is then straightforward and is presented mainly in Chapters 7 and 8. Applications and more advanced topics are presented in Chapters 9 and 10.

1.2 CLASSIFICATION OF NUMERICAL METHODS

Engineering models are used to establish a relationship between the input (source, stimulus) to a system and the output (response) from the system. This is shown schematically in Fig. 1.1. The process of Conceptualizing and formulation leads to the expression of the physical laws describing the system in a form of the type:

$$\mathcal{L}(\Phi) = \alpha \tag{1.1}$$

where

$\mathcal{L}$ = a mathematical operator

Φ = a field function that must be determined

α = a source function

Basic classification schemes for numerical methods can be arrived at by examining the nature of Equation (1.1).

One criterion for classification is the domain in which the operator, the field, and source functions are defined. If these are defined in the time-domain, then the method is described as a time-domain (TD) method.

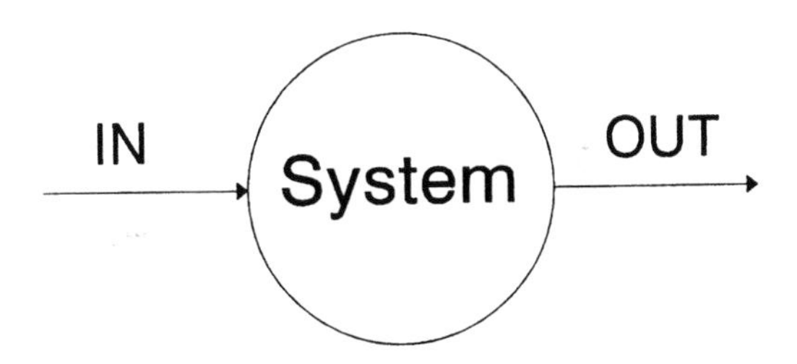

Fig. 1.1 Input/output in a system

Alternatively, the frequency domain may be chosen, leading to frequency-domain (FD) methods. Examples of the two formulations for the circuit shown in Fig. 1.2 are:

$$V_0 \cos \omega t = i(t) R + L \frac{di(t)}{dt}$$

for TD formulation and

$$V_0 = \bar{I}(R + j\omega L)$$

for FD formulation, where $\bar{I}$ represents the current phasor.

Clearly, the TD formulation is suitable for studying transients and nonlinear phenomena, while the FD formulation is straightforward for studying the steady-state response to a sinusoidal excitation. In the TD, the function fully characterizing the system is its impulse response $h(t)$, while in the FD, the frequency response $H(j\omega)$ offers a complete system description. Since $h(t)$ and $H(j\omega)$ form a Fourier transform pair, any information available in the TD can be converted into the FD, and vice-versa. However, although the two descriptions are formally equivalent, issues of efficiency normally dictate which approach is the most convenient in a particular problem. For example, if the steady-state response at a single frequency is required, the natural choice must be a FD method. For tran-

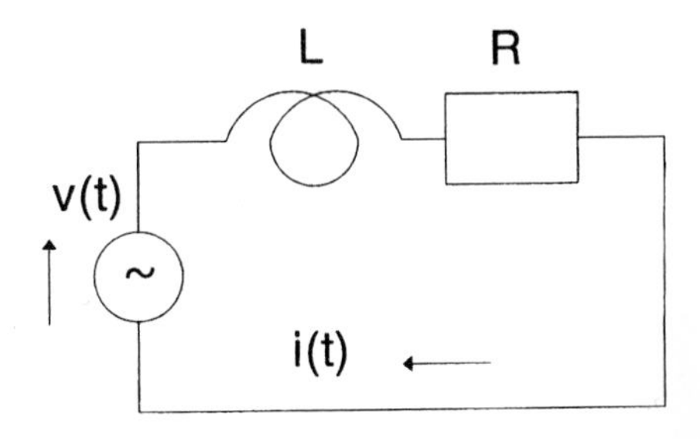

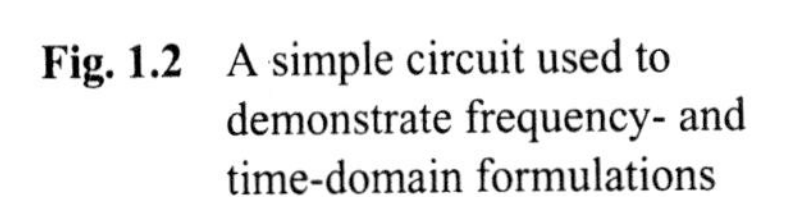
Fig. 1.2 A simple circuit used to demonstrate frequency- and time-domain formulations

sients, or when the response over a wide frequency range is required, a TD method may be the most appropriate.

For the simple example shown in Fig. 1.2, both "transfer functions" $h(t)$ and $H(j\omega)$ can be readily obtained in an analytical form. However, in most practical problems, the transfer function is a multivariable, multidimensional, discrete function defined over a large number of mesh points. It occupies a large amount of computer memory, and its determination and manipulation forms the formal task of solving a problem by numerical simulation.

Another criterion for classification is the nature of the operator $\mathcal{L}$ used in Equation (1.1). It may be expressed in a differential or in an integral form, thus resulting into differential (DE) and integral (IE) numerical modeling methods, respectively. An example of differential and integral formulations of the same basic physical ideas can be seen in Gauss's Law:

$$\nabla \cdot \mathbf{D} = \rho$$

$$\int_S \mathbf{D}\, d\mathbf{S} = Q$$

The former equation is a differential formulation and must be enforced at every point in the problem space. The latter is an integral formulation and must be enforced on surfaces in the problem space. Numerical methods fall broadly within these two categories.

A further class of methods, with its own special characteristics, is that of ray methods, which are based on concepts borrowed from optics. Several formulations are also available combining more than one method, and these are described as hybrid numerical methods.

In formulating a problem for solution by digital computer, the convenient concepts used in analytical techniques of infinitesimally small time and space steps (dt and dx, respectively) must be replaced by small but finite steps (Δt and Δx, respectively). This is fundamental in computer simulation because, if time-steps were infinitesimally small, a computer with a finite clock speed would require an infinitely long time to do the simplest calculation. Similarly, an infinitesimally small space-step would require a computer with an infinite number of memory locations. Therefore, implicit in every numerical simulation is the construction of a grid or mesh of points, covering the entire problem space, on which the relevant physical laws are enforced at successive time steps. It is arguable that at the atomic level physical processes are finite. However, in most practical simulations, the replacement of the concept of a continuum by the discrete approach is dictated by practical considerations, such as computer storage and run-time. Discretization in time and space therefore introduces errors that must be quantified and controlled in every numerical simulation.

In a differential method, the entire space has to be discretized, and this normally results in large computational requirements. In practice, if a suitable physical outer boundary surface does not exist in the problem (such as in the case of open-boundary problems), then an artificial "numerical boundary" must be defined to contain the computation within manageable limits. Defining the correct boundary conditions on numerical boundaries is not an easy task. On the positive side, the enforcement of physical laws on all points in space means that fine features, irregular shapes, and material inhomogeneities can be easily dealt with. In addition, in spite of the large number of quantities to be determined on grid points in space, the resulting equations can be solved relatively easily.

In contrast, in an integral method, physical laws are enforced on grid points lying on important surfaces of the problems and, hence, the number of quantities to be determined is smaller. However, the equations to be solved are usually more complex, so a relatively small number of points can be handled. Open-boundary problems can be dealt with rigorously in IE formulations.

This brief outline of DE and IE methods indicates that each general class has its own advantages and disadvantages, and that the best method is application dependent. A more detailed discussion of numerical methods may be found in [5].

The need for discretization in numerical simulation leads to the question of the choice of appropriate time- and space-steps. An answer to this question cannot be given without reference to the method used, errors, and application requirements. However, a useful practical rule is to choose a space discretization length that is smaller than one-tenth of the smallest wavelength of interest. For both classes of finite methods (DE and IE), and with computing facilities commonly available, it is difficult to envisage solving problems in three dimensions which are larger in physical size than a few wavelengths. At very high frequencies, it is advantageous to use ray methods. These methods can be applied when the wavelength λ is much smaller than the size of the features being modeled. Fields are calculated by taking into account reflected and diffracted rays. Methods based on geometrical optics (GO), the geometrical theory of diffraction (GTD), and other more general approaches are available.

1.3 ELECTRICAL CIRCUIT ANALOGS OF PHYSICAL SYSTEMS

It is undoubtedly true that humans model best when they use a medium with which they are familiar. Mechanical phenomena are closest to human experience and the first to be understood and formulated in scientific terms. The field of mechanics has long been understood in terms of a coherent self-consistent set of scientific principles. It was therefore natural that, during the beginnings of electrical science, mechanical models were used to aid understanding and make predictions. Electrical science now has the status of a well established scientific theory, and it therefore can be used as a model for studying other phenomena. While mathematicians are familiar with differential equations, electrical and electronic engineers have an intuitive understanding of how electrical circuits work and the significance of each circuit component. They are, therefore, more comfortable with circuit models than with the more abstract mathematical models normally used to study electromagnetic fields. Circuit models can also be used to study thermal and mechanical phenomena—an interesting development, considering the situation at the turn of the century. The purpose of this book is to provide a systematic treatment of the modeling process based on electrical circuit analogs. In this chapter, however, only a general introduction will be given. The modeling principles will be outlined with a broad brush to assist the reader in following more comfortably the detailed treatment in subsequent chapters.

The idea of using electrical circuits to model fields is not new. More recently, the work of Kron and others signposted the equivalence between field and circuit ideas [6–8]. However, no substantial progress could be made in exploiting these ideas, because the circuit models could not be solved using the calculation tools available at the time. Further development had to wait for the introduction of modern digital computers and the pioneering work of Johns and Beurle [9], which provided sufficient impetus for a rapid advance.

The analogy between circuits and fields can be understood by considering the transmission line circuit shown in Fig. 1.3. The voltage v and current i on the line are functions of time t and distance x. Kirchhoff's voltage (KVL) and current (KCL) laws on the lines give:

$$-\frac{\partial v}{\partial x}\Delta x = L\frac{\partial i}{\partial t} \tag{1.2}$$

$$-\frac{\partial i}{\partial x}\Delta x = C\frac{\partial v}{\partial t} + \frac{v}{R} \tag{1.3}$$

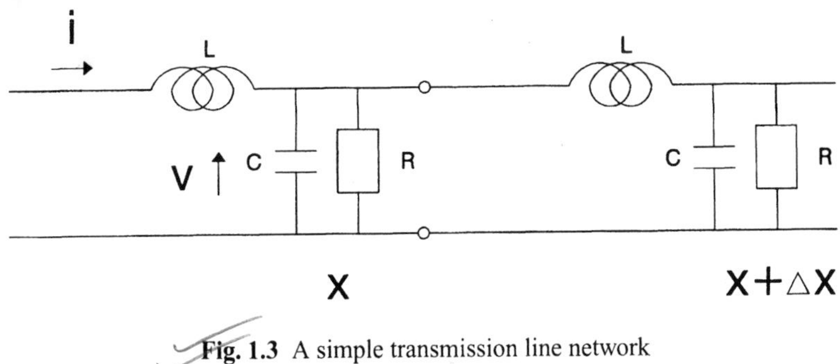

Fig. 1.3 A simple transmission line network

These expressions are strictly correct provided that $\Delta x \to 0$. Equations (1.2) and (1.3) may be manipulated to eliminate v and thus provide an equation containing i only:

$$\frac{\partial^2 i}{\partial x^2} = \frac{LC}{(\Delta x)^2} \frac{\partial^2 i}{\partial t^2} + \frac{L}{(\Delta x)^2 R} \frac{\partial i}{\partial t} \tag{1.4}$$

It can also be shown that in a one-dimensional EM field problem, the current density j is determined by the following equation [10]:

$$\frac{\partial^2 j}{\partial x^2} = \mu\varepsilon \frac{\partial^2 j}{\partial t^2} + \mu\sigma \frac{\partial j}{\partial t} \tag{1.5}$$

where μ, ε, and σ are the magnetic permeability, electric permittivity, and electrical conductivity of the medium.

Field components E, B are also described by equations similar to Equation (1.5), which is known as the wave equation in a lossy medium. The analogy between the circuit [Equation (1.4)] and field [Equation (1.5)] problems is based on the similar form of these two equations. Similarly, analogues of Equations (1.4) and (1.5) may be found in three-dimensional problems, as shown in Chapter 6. The laws governing circuit behavior have the same syntactical structure as those governing field behavior. The isomorphism between Equations (1.4) and (1.5) means that the behavior of fields may be understood by studying the behavior of circuits. For the particular example considered here, the equivalence between circuits and fields is summarized in Table 1.1.

It is worth considering further the nature of Equation (1.5). Assuming a harmonic variation for j, i.e., $j(x, t) = j_0 \cos(\omega t - \beta x)$, the two terms on the right-hand side of this equation can be evaluated and their

Table 1.1 Equivalence between circuits and fields

Circuit		*EM Field*
i	⇔	j
$L/\Delta x$	⇔	μ
$C/\Delta x$	⇔	ε
$1/R\Delta x$	⇔	σ

magnitudes compared. The first term describes wave-like behavior, and given that $\partial j/\partial t \sim \omega j$, and $\partial^2 j/\partial t^2 \sim \omega^2 j$,

$$|\text{wave term}| \sim \mu\varepsilon\omega^2$$

The second term describes diffusion-like behavior, and its magnitude is

$$|\text{diffusion term}| \sim \mu\sigma\omega$$

Hence,

$$\frac{|\text{wave term}|}{|\text{diffusion term}|} = \frac{\omega\varepsilon}{\sigma}$$

In cases where $\omega\varepsilon >> \sigma$, wave behavior dominates. This is the case with propagation in air and low-loss dielectrics at high frequencies. When $\omega\varepsilon < \sigma$, diffusion behavior prevails, as in propagation at low frequencies in lossy media. The circuit equation (1.4) thus can be used to model waves, diffusion, and any combination of the two. To model diffusion, the first term on the left-hand side of this equation must be negligible compared to the second. The diffusion-dominated equation (1.4) can then be used to model thermal conduction in a material. In this case, the temperature distribution $\theta(x,t)$ is determined by the diffusion equation:

$$\frac{\partial^2\theta}{\partial x^2} = \frac{S}{k_{th}}\frac{\partial\theta}{\partial t} \tag{1.6}$$

where

S = the specific heat in J/Km^3

k_{th} = the thermal conductivity in W/Km

Clearly, the analogy between Equation (1.6) and the diffusion-dominated Equation (1.4) requires the equivalence shown in Table 1.2.

Table 1.2 Circuit and thermal equivalence

Circuit		*Thermal*
$\frac{L}{\Delta x}$	$\Leftrightarrow$	S
$R \cdot \Delta x$	$\Leftrightarrow$	k_{th}

Other circuits can also be used to model thermal problems, as will be discussed in more detail further in this book. The purpose of the previous brief treatment is merely to illustrate the principles of establishing analogies between different physical problems.

The circuit analogs described above would be of only educational value if their solution could not be found in a relatively simple manner. In engineering practice, it is important to develop models that clearly illustrate the important physical interaction, but a solution of these models is also necessary to predict behavior and optimize designs. Circuit models such as the one shown in Fig. 1.3 can be very complex in practice, especially when they are generalized to describe two- and three-dimensional distributions, as it will be shown later.

TLM provides a systematic, elegant, and efficient procedure for solving these networks. It is based on using transmission line elements to describe all energy storage elements. It is therefore important to summarize (see Chapter 2) aspects of transmission line theory that are important in understanding the implementation of TLM.

REFERENCES

[1] Lodge, O. 1889. *Modern Views of Electricity.* London: Macmillan.

[2] Hempel, C.G. 1965. *Aspects of Scientific Exploration and other Essays in the Philosophy of Science.* London: Macmillan.

[3] Hammond, P. 1988. Some thoughts on the numerical modelling of electromagnetic processes. *International Journal of Numerical Modelling* 1, 3–6.

[4] Johns, P.B. 1979. The art of modelling. *Electronics and Power,* 565–569.

[5] Miller, E.K. 1988. A selective survey of computational electromagnetics. IEEE Trans. AP-26, 1281–1305.

[6] Kron, G. 1944. Equivalent circuit of the field equations of Maxwell. *Proc IRE* 32, 289–299.

[7] Whinnery, J.R., and S. Ramo. 1944. A new approach to the solution of high-frequency problems. *Proc. IRE* 32, 284–288.

[8] Whinnery, J.R., et al. 1944. Network analyser studies of electromagnetic cavity resonators. *Proc IRE,* 32, 360–367.

[9] Johns, P.B., and R.L. Beurle. 1971. Numerical solution of two-dimensional scattering problems using transmission-line matrix. *Proc IEE* 118, 1203–1208.

[10] Christopoulos, C. 1990. *An Introduction to Applied Electromagnetism.* New York: John Wiley & Sons.

2

Transmission Line Theory

A lossless transmission line is described by a distributed inductance and capacitance, as shown schematically in Fig. 2.1. The inductance and capacitance per unit length are designated by L_d and C_d, respectively. Transmission line problems can be posed in one of two different ways.

First, one end of the line may be suddenly connected to a source, and the voltage and current are then required anywhere on the line (including the termination) as a function of time. This is the transient response of the line and it requires a model of the line in the time domain.

Second, one end of the line may be connected to a sinusoidal source, and the voltage and current anywhere on the line are required after a steady-state has been established. This problem is best addressed by constructing a line model in the frequency domain. These two model formulations are studied separately in the next two sections.

2.1 TRANSIENT RESPONSE OF A LINE

Let us consider a lossless line suddenly connected to a dc source as shown in Fig. 2.2. The conditions on the line at a time Δt after the connection of

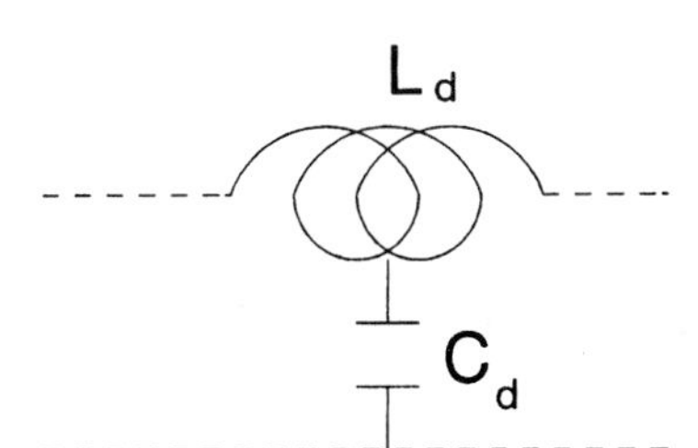

Fig. 2.1 Schematic of a transmission line section

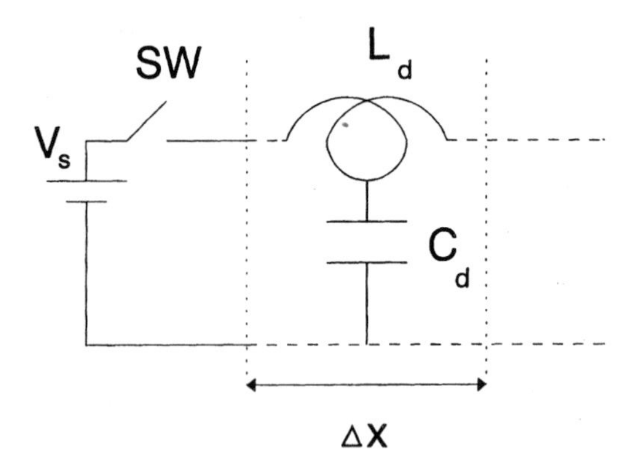

Fig. 2.2 Charging transient on a line

the source, when the source voltage has penetrated a distance Δx into the line, are as described below.

The capacitance of the energized segment of the line $C_d\ \Delta x$ has charged to a voltage V_s and hence a charge $\Delta Q = C_d\ \Delta x\ V_s$ has been transferred. The current flowing into the line is:

$$i = \frac{\Delta Q}{\Delta t} = C_d V_s \frac{\Delta x}{\Delta t} = C_d V_s u$$

where u is the velocity of propagation along the line of the disturbance caused by the connection of the source. The flow of this current will establish a magnetic flux Φ associated with the line inductance.

$$\Phi = L_d \Delta x\ i = L_d \Delta x C_d V_s u$$

From Faraday's Law, the rate of change of flux must be equal to the line voltage, i.e.,

$$V_s = \frac{\Delta \Phi}{\Delta t} = L_d C_d V_s u^2$$

Hence, the velocity of propagation on the line is:

$$u = \frac{1}{\sqrt{L_d C_d}} \text{ m/s} \tag{2.1}$$

Substituting (2.1) into the equation for the current gives:

$$i = C_d V_s \frac{1}{\sqrt{L_d C_d}} = \frac{V_s}{\sqrt{L_d / C_d}}$$

The quantity in the denominator of this expression is described as the "characteristic" or "surge" impedance of the line.

$$Z_0 = \sqrt{\frac{L_d}{C_d}} \ \Omega \tag{2.2}$$

Whenever a voltage pulse V is propagating along the line, there is an associated current pulse, $I = V/Z_0$. Therefore, until conditions on the line are affected by discontinuities (e.g., terminations), the input impedance of the line is Z_0. The voltage and current pulses are shown schematically in Fig. 2.3a. When studying pulses traveling on transmission lines, the voltage and current positive reference directions must be specified, and the direction of propagation u must also be indicated. These are shown in Fig. 2.3a. For the plotting convention adopted, both the voltage and current pulses are positive.

Let us now consider the moment when the voltage and current pulses travel the length, ℓ, of the line and encounter an open-circuit termination. The current at the open-circuit must be zero, so a current pulse must be initiated at the termination traveling toward the source of magnitude $-V_s/Z_0$ so that the line current is progressively forced to zero. This is

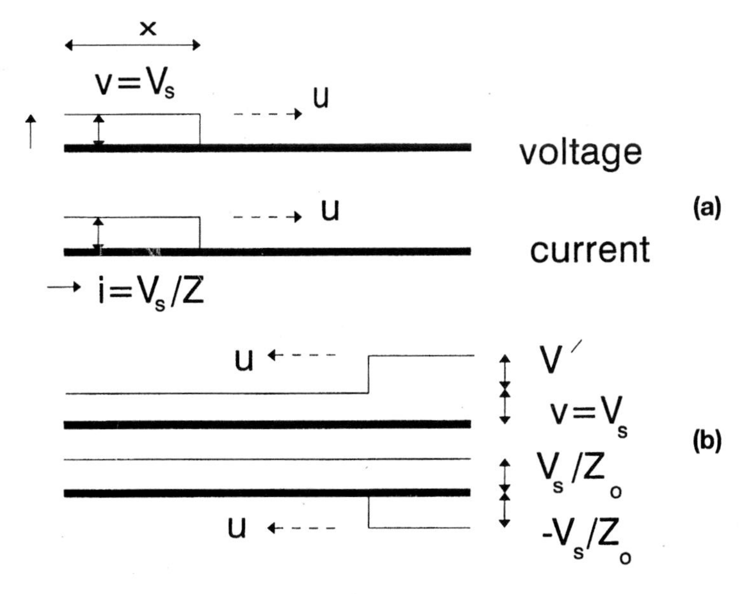

Fig. 2.3 Propagation of an electrical disturbance on a line, (a) before and (b) after reflection from the termination. Solid arrows indicate positive reference directions for voltage and current; broken arrows indicate propagation directions.

shown schematically in Fig. 2.3b, with the associated voltage pulse of a magnitude V' yet undefined. When these pulses reflected from the open-circuit termination arrive at the source end, the voltage along the entire length of the line is $V' + V_s$, and the current is zero. This condition occurs after a time 2τ following the connection of the source, where τ is the transit time of the line and is equal to ℓ/u.

The magnitude of pulse V' can be obtained by demanding energy conservation. At the end of the 2τ period, the energy stored on the line is associated with the capacitance only

$$\frac{1}{2}(C_d \cdot l)\,(V_s + V')^2$$

The energy supplied by the source is:

$$V_s \frac{V_s}{Z_0} 2\tau$$

Equating these two expressions and substituting Z_0 and τ in terms of the line parameters gives $V' = V_s$.

Hence, a voltage pulse incident on an open circuit termination is reflected so that the total voltage at the termination is equal to twice the voltage associated with the incident pulse. To distinguish between incident and reflected pulses, superscripts i and r are used, respectively. Hence, for the example just described, the incident and reflected pulses at the open-circuit load are:

$$V^i = V_s \qquad V^r = V_s$$

$$I^i = V_s/Z_0 \qquad I^r = -V_s/Z_0$$

A useful corollary of these results is that an observer at the end of a line toward which a pulse V^i is propagating can replace the line by a Thevenin equivalent circuit, where the voltage source is equal to the open circuit voltage, which was shown to be $2V^i$, and an impedance equal to the input impedance of the line, which earlier was shown to be Z_0. The line and its equivalent circuit are shown schematically in Fig. 2.4a and 2.4b. This procedure is valid irrespective of the actual line termination, provided that V^i does not change due to reflections from other parts of the network to which the line is connected. This, in turn, implies that the Thevenin equivalent circuit is valid for a limited time period (the transit time τ of the line), and it has to be updated whenever new incident pulses arrive.

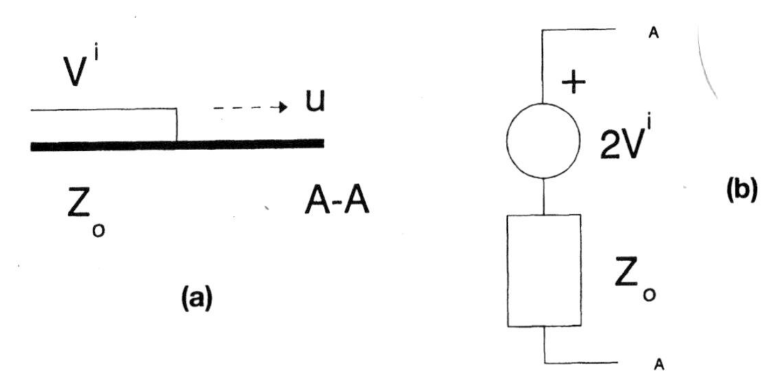

Fig. 2.4 Thevenin equivalent circuit seen at A–A

As an example of the application of the Thevenin equivalent circuit, the voltage on a line terminated with a resistance R is calculated. It is assumed that voltage and current pulses V^i and V^i/Z_0, respectively, are incident at the termination. Replacing the line by its Thevenin equivalent and connecting the load to it as shown in Fig. 2.5 gives the voltage:

$$V = \frac{R}{R + Z_0} 2V^i$$

The voltage reflected and traveling away from the termination is

$$V^r = V - V^i = \left(\frac{R - Z_0}{R + Z_0}\right) V^i$$

The quantity in the brackets is called the reflection coefficient

$$\Gamma = \frac{R - Z_0}{R + Z_0} \tag{2.3}$$

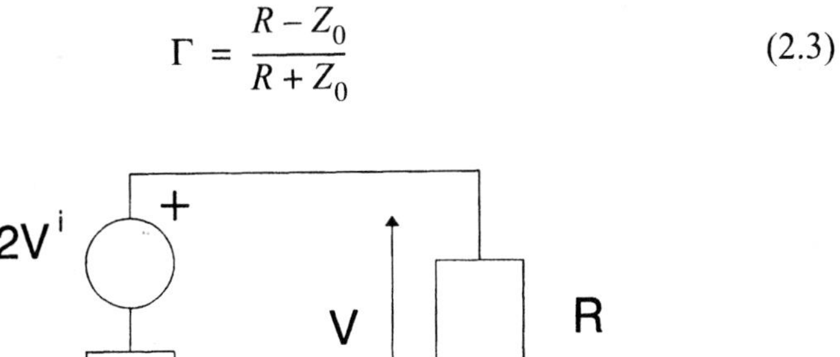

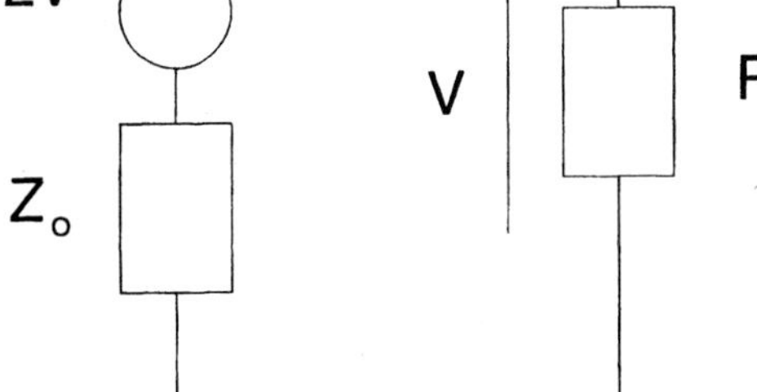

Fig. 2.5 Conditions at the end of a line terminated by a resistance

For an open-circuit termination ($R \to \infty$) $\Gamma = 1$, for a short-circuit ($R \to 0$) $\Gamma = -1$, and for the special case of the termination being equal to the characteristic impedance of the line ($R = Z_0$) $\Gamma = 0$. In the latter case of no reflections, the line is described as being matched by its own characteristic impedance.

In a complex system consisting of several lines, each of which differs in length, characteristic impedance, and termination method, a great many reflections take place, and the voltage or current anywhere in the network consists of a superposition of these reflected signals. These phenomena may be studied in a systematic way using the Bewley Lattice technique. Further details may be found in References [1] and [2].

2.2 SINUSOIDAL STEADY-STATE RESPONSE OF A LINE

The approach described in Section 2.1 is generally applicable and could be used to study the response of the line to a sinusoidal excitation. It would, however, be necessary to determine the large number of reflections leading up to the establishment of a steady state. If only the steady state is required, a simpler approach is possible, and this is described here.

The equivalent circuit of a lossless line is shown in Fig. 2.6, where $\bar{V}(x)$ and $\bar{I}(x)$ are the voltage and current phasors. The objective is to determine how these phasors vary along the length of the line. The instantaneous voltage can then be obtained from the expression

$$v(x, t) = \Re e\,[\bar{V}(x)\,e^{j\omega t}]$$

where ω is the angular frequency of the harmonic signal.

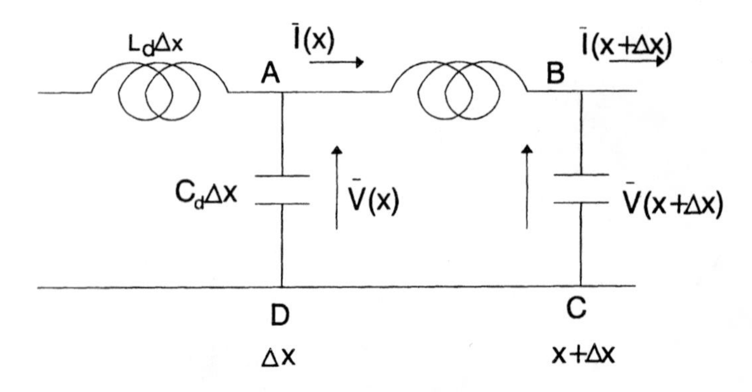

Fig. 2.6 Phasor quantities on a line

Applying KVL in mesh ABCD gives

$$\bar{V}(x) - \bar{V}(x+\Delta x) = j\omega L_d \Delta x \bar{I}(x)$$

Hence, as $\Delta x \to 0$,

$$-\frac{d\bar{V}(x)}{dx} = j\omega L_d \bar{I}(x) \tag{2.4}$$

Similarly, applying KCL at node B gives

$$\bar{I}(x) - \bar{I}(x+\Delta x) = j\omega C_d \Delta x \bar{V}(x+\Delta x)$$

Hence,

$$-\frac{\bar{I}(x+\Delta x) - I(x)}{\Delta x} = j\omega C_d \bar{V}(x+\Delta x)$$

Expanding in a Taylor series gives

$$\bar{I}(x+\Delta x) = I(x) + \Delta x \frac{d\bar{I}(x)}{dx} + \ldots$$

In the limit as $\Delta x \to 0$,

$$I(x+\Delta x) \to I(x)$$

and

$$\frac{I(x+\Delta x) - I(x)}{\Delta x} \to \frac{dI(x)}{dx}$$

Similarly, $V(x+\Delta x) \to V(x)$. Therefore, KCL reduces to

$$-\frac{dI(x)}{dx} = j\omega C_d \bar{V}(x) \tag{2.5}$$

It should be noted that Equations (2.4) and (2.5) are strictly correct when $\Delta x \to 0$. For a finite value of Δx, these expressions are only approximate.

Taking the derivative of Equation (2.4) with respect to x and substituting for $\bar{I}(x)$ from Equation (2.5) gives

$$\frac{d^2\bar{V}(x)}{dx^2} = -\omega^2 L_d C_d \bar{V}(x) = -\beta^2 \bar{V}(x)$$

where β is the phase constant of the line. A similar expression is obtained for $\bar{I}(x)$, giving solutions of the general form

$$\bar{V}(x) = V_A e^{j\beta x} + V_B e^{-j\beta x}$$

$$I(x) = I_A e^{j\beta x} + I_B e^{-j\beta x} \tag{2.6}$$

where V_A, V_B, I_A, and I_B are complex constants to be determined from the boundary conditions (voltage/current at the line terminations).

It is worth exploring further the nature of the terms shown in Equation (2.6). From the voltage expression,

$$v(x,t) = \Re_e[\bar{V}(x)e^{j\omega t}] = \Re_e\left[V_A e^{j(\omega t + \beta x)} + V_B e^{j(\omega t - \beta x)}\right]$$

The first term represents wave-like behavior where a point of constant phase must be constrained by the expression

$$\omega t + \beta x = \text{constant}$$

Hence, it follows that $\omega dt + \beta dx = 0$. The phase velocity for this wave therefore is

$$u = \frac{dx}{dt} = -\frac{\omega}{\beta} = -\frac{\omega}{\omega\sqrt{L_d C_d}} = -\frac{1}{\sqrt{L_d C_d}}$$

Similarly, the second term represents waves with velocity

$$u = +\frac{1}{\sqrt{L_d C_d}}$$

The conclusion from this analysis is that the response of the line in steady state consists of the superposition of waves traveling in the forward $+x$ and backward $-x$ directions. The voltage and current waves traveling in the forward direction $V_B e^{-j\beta x}$ and $I_B e^{-j\beta x}$, respectively, are not independent of each other. Substituting these two expressions in both sides of Equation (2.4) gives:

$$I_B = \frac{V_B}{\sqrt{L_d/C_d}} = \frac{V_B}{Z_0}$$

where Z_0 is the characteristic impedance of the line defined in Equation (2.2).

Similarly, for the backward traveling wave,

$$I_A = -\frac{V_A}{\sqrt{L_d/C_d}} = -\frac{V_A}{Z_0}$$

Equations (2.6) thus reduce to

$$\bar{V}(x) = V_A e^{j\beta x} + V_B e^{-j\beta x}$$

$$I(x) = -\frac{V_A}{Z_0} e^{+j\beta x} + \frac{V_B}{Z_0} e^{-j\beta x} \tag{2.7}$$

The constants V_A, V_B can be obtained from the voltage and current at the start $x = 0$ of the line $\bar{V}(o)$ and $\bar{I}(o)$, respectively.

Substituting x = 0 into Equation (2.7) gives

$$V_A = \frac{\bar{V}(o) - I(o) Z_0}{2}$$

$$V_B = \frac{\bar{V}(o) + I(o) Z_0}{2}$$

Thus, after some rearrangement, Equation (2.7) may be expressed as

$$\bar{V}(x) = \bar{V}(o) \cos(\beta x) - I(o) j Z_0 \sin(\beta x)$$

$$I(x) = \bar{V}(o) \frac{-j \sin(\beta x)}{Z_0} + I(o) \cos(\beta x) \tag{2.8}$$

Equation (2.8) expresses the voltage and current phasors as functions of the phasors at the start of the line. The voltage and current at the end of the line $x = \ell$, may also be related to conditions at the start:

$$\begin{bmatrix} \bar{V}(o) \\ I(o) \end{bmatrix} = \begin{bmatrix} \cos(\beta\ell) & jZ_0 \sin(\beta\ell) \\ j\frac{\sin(\beta\ell)}{Z_0} & \cos(\beta\ell) \end{bmatrix} \begin{bmatrix} \bar{V}(\ell) \\ I(\ell) \end{bmatrix}$$

or

$$\begin{bmatrix} \bar{V}(o) \\ I(o) \end{bmatrix} = \begin{bmatrix} A & B \\ C & D \end{bmatrix} \begin{bmatrix} \bar{V}(1) \\ I(1) \end{bmatrix} \tag{2.9}$$

where A, B, C, and D are the elements of the transmission matrix $[T]$ of the line.

The input impedance of the line is

$$Z_{in} = \frac{\bar{V}(o)}{I(o)} = \frac{AZ_\ell + B}{CZ_\ell + D}$$

where Z_ℓ is the load at the termination of the line. Substituting for the transmission parameters from Equation (2.9) gives

$$\frac{Z_{in}}{Z_o} = \frac{\dfrac{Z_\ell}{Z_0} + j\tan(\beta\ell)}{1 + j\dfrac{Z_\ell}{Z_0}\tan(\beta\ell)} \tag{2.10}$$

Clearly, the line input impedance varies widely, ranging from large capacitive to large inductive values, depending on the line parameters and the frequency. Further details may be found in Ref. [2].

2.3 DISPERSIVE EFFECTS IN DISCRETIZED TRANSMISSION LINE MODELS

In the process of discretizing a continuous system, errors are introduced which become apparent at high frequencies. The nature of these errors may be understood by reference to the derivation of the system equations in the last two sections, where it was assumed that in the limit $\Delta x \to 0$

$$v(x + \Delta x, t) \approx v(x, t) + \frac{\partial v(x, t)}{\partial x}\Delta x \to v(x, t)$$

This approximation is acceptable provided that

$$\left|\frac{\partial v(x, t)}{\partial x}\Delta x\right| \ll |v(x)| \tag{2.11}$$

Let us assume a wave-like behavior for $v(x,t)$; i.e.,

$$v(x, t) = V_o \sin(\omega t - kx) \tag{2.12}$$

Then the magnitudes of the two terms in Equation (2.11) are

$$\left|\frac{\partial v}{\partial x}\Delta x\right| = \frac{2\pi}{\lambda}\Delta x V_o$$

$$|v| = V_o$$

Hence, the approximation in Equation (2.11) is acceptable, provided that

$$\Delta x \ll \lambda \tag{2.13}$$

The significance of this expression is that it sets a limit to the value of the space discretization length, which depends on the shortest wavelength of interest in the computation. A commonly adopted rule of thumb is to select x smaller than one-tenth of the smallest wavelength.

It is possible to obtain a quantitative measure of the errors introduced by discretization. It is found that propagation in the circuit shown in Fig. 2.6, where x is finite, is dispersive; i.e., the velocity of propagation depends on the frequency [3]. The dispersion relation may be obtained as follows. Applying KVL in the two loops between x – Δx, x and x, x + Δx gives

$$v(x-\Delta x) = L\frac{\partial}{\partial t}i(x-\Delta x) + v(x) \tag{2.14}$$

$$v(x) = L\frac{\partial}{\partial t}i(x) + v(x+\Delta x) \tag{2.15}$$

where the dependence of v and i on time is not shown. Similarly, KCL at the node at x gives

$$i(x-\Delta x) = C\frac{\partial}{\partial t}v(x) + i(x) \tag{2.16}$$

Subtracting Equation (2.15) from (2.14) and combining with (2.16) to eliminate current dependence gives

$$LC\frac{\partial^2}{\partial t^2}v(x) = v(x+\Delta x) + v(x-\Delta x) - 2v(x) \tag{2.17}$$

This is an accurate difference equation describing the discrete system. Assuming a wave-like dependence for all voltages as shown in Equation (2.12) and substituting in (2.17) gives (after some algebra)

$$\omega^2 = \frac{1}{LC}4\sin^2\left(\frac{k\Delta x}{2}\right) \tag{2.18}$$

This is the dispersion relation for propagation on the line, relating frequency to wavelength. The wave propagation velocity is

$$u_w = \frac{\omega}{k} = \frac{\Delta x}{\sqrt{LC}} \left[\frac{\sin\left(\frac{k\Delta x}{2}\right)}{\left(\frac{k\Delta x}{2}\right)} \right] \tag{2.19}$$

The quantity $\Delta x / \sqrt{LC}$ is the velocity of propagation u_0 on the distributed line. It is clear from Equation (2.19) that propagation at u_0 is only possible at low frequencies, when the term in the brackets tends to one. For finite values of Δx, some deviation from u_0 always occurs. As an example, if Δx is chosen to be equal to $\lambda/10$, the velocity of propagation is $u_w = 0.984\ u_0$. This calculation gives a measure of the velocity error (≈ 2 percent) that must be expected for this discretization.

REFERENCES

[1] Greenwood, A. 1971. *Electrical Transients in Power Systems.* New York: John Wiley & Sons.

[2] Ramo, S., et al. 1984. *Fields and Waves in Communication Electronics,* Ch. 5. New York: John Wiley & Sons.

[3] Hirose, A., and E.E. Lonngren. 1985. *An Introduction to Wave Phenomena.* New York: John Wiley & Sons.

3

Discrete Models of Lumped Components

As discussed in Section 1.3, many physical systems may be modeled by electrical circuit analogs. These consist of lumped components such as inductors and capacitors. Even systems with distributed components, such as transmission lines, may be rendered discrete in space by dividing them into small sections of length Δx. Each section is then described by lumping together its capacitance and inductance. The lumped component representation of a distributed system is tantamount to space discretization. This is the first step in obtaining a numerical solution to any problem, as it makes finite the number of problem unknowns and, therefore, computer storage locations.

A second step is necessary before the problem is ready for computation. This is discretization in time, and the purpose of this chapter is to explain this process. The discretized models of capacitors and inductors are developed first.

3.1 "LINK" AND "STUB" MODELS OF CAPACITORS

Let us consider capacitor C shown in Fig. 3.1a. Intuitively, it appears possible to develop a discrete model of this capacitor based on a transmission line segment or "link," as shown in Fig. 3.1b. Let us see what the parameters of such a model should be. The capacitance and inductance per unit length for the link line are chosen to be C_d and L_d, and the length and propagation delay are $\Delta\ell$ and Δt, respectively. Clearly, to model a capacitance C, C_d must be chosen so that

$$C_d \Delta\ell = C$$

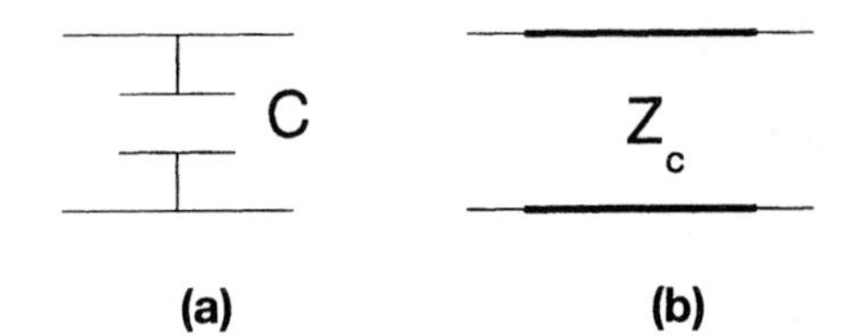

Fig. 3.1 (a) a capacitor and (b) its link model

The velocity of propagation along the line is then

$$u = \frac{\Delta \ell}{\Delta t} = \frac{1}{\sqrt{L_d C_d}}$$

Therefore,

$$L_d = \left(\frac{\Delta t}{\Delta \ell}\right)^2 \frac{1}{C_d}$$

The characteristic impedance of the line is then

$$Z_c = \sqrt{\frac{L_d}{C_d}} = \frac{\Delta t}{C} \tag{3.1}$$

A line with an impedance Z_C appears as a capacitor C, but with an associated inductance equal to

$$L_e = L_d \Delta \ell = \frac{(\Delta t)^2}{C} \tag{3.2}$$

The inductance L_e should be regarded as a modeling error and could be minimized by decreasing Δt. Under practical conditions, no component is purely capacitive. It may be argued that the error inductance L_e is none other than the stray inductance associated with a real capacitor, C. This feature of the model may be exploited to give a more realistic representation of real components.

In the link model of the capacitor, voltage, and current can be updated only at fixed times—multiples of Δt. Hence, the problem is discretized in time with Δt being the discretization time, or the time-step of the computation.

An alternative model, known as a "stub" model, is also possible, as shown in Fig 3.2. The stub line shown has parameters C_d, L_d, and length $\Delta \ell$, but the round-trip time is Δt. The stub is terminated on an open circuit. This is a reasonable choice, as it emphasizes voltage differences, storage in the electric field and, hence, mainly capacitive behavior. Following a procedure similar to that for the link model gives

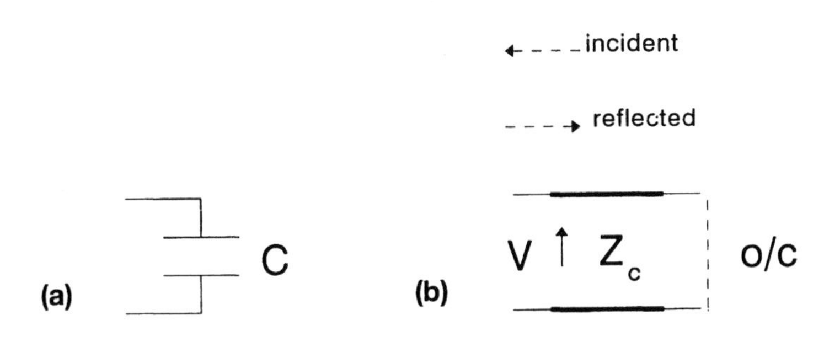

Fig. 3.2 (a) a capacitor and (b) its link model

$$C_d \Delta \ell = C$$

$$u = \frac{\Delta \ell}{\Delta t/2} = \frac{1}{\sqrt{L_d C_d}} \text{ and, hence,}$$

$$L_d = \frac{(\Delta t)^2}{4C\Delta \ell}$$

The characteristic impedance of the stub capacitor model is

$$Z_c = \sqrt{\frac{L_d}{C_d}} = \frac{\Delta t}{2C} \tag{3.3}$$

The associated error inductance is

$$L_e = L_d \cdot \Delta \ell = \frac{(\Delta t)^2}{4C} \tag{3.4}$$

The same comments regarding the error inductance apply as for the link model.

Another way of looking at the accuracy of the model is to examine the input impedance to the stub. Using Equation (2.10) and substituting $Z_\ell = \infty$(because the stub termination is an open circuit) gives

$$Z_{in} = \frac{Z_c}{j\tan(\beta \ell)}$$

where

$$\beta \ell = \frac{\omega}{u} \ell = \omega \frac{\Delta t}{2}$$

Hence, the input admittance is

$$Y_{in} = \frac{j\tan\left(\frac{\omega\Delta t}{2}\right)}{Z_c} = j\frac{2C}{\Delta t}\left[\frac{\omega\Delta t}{2} + \frac{1}{3}\left(\frac{\omega\Delta t}{2}\right)^3 + \dots\right]$$

$$= j\omega C + j\frac{C\omega^3 (\Delta t)^2}{12} + \dots$$

The error in the admittance is of the order of $(\Delta t)^2$ and decreases as the time-step is reduced. In both the link and stub transmission line models (TLM), the modeling errors can be described clearly in terms of electrical components (i.e., error or parasitic inductance). It is therefore easier for the applied scientist to ascertain whether a certain error is acceptable. Following this initial modeling error, the model is solved exactly, and no further approximations or errors are introduced. Since the model of a capacitor results in a passive circuit which is solved exactly, stability is beyond doubt.

Let us now look at an example of how a stub model of a capacitor can be used to solve the simple network shown in Fig. 3.3a.

The TLM model of this network using a stub capacitor model is shown in Fig 3.3b, where voltages and current are indicated at time step $k\Delta t$. The voltage incident at the stub port is indicated by ${}_kV_c^i$. The characteristic impedance of the stub Z_c is obtained from Equation (3.3) where Δt is the round-trip time for the stub. The correct choice of the value of Δt will be discussed after the basic solution algorithm has been explained. The next step is a further simplification of the TLM network obtained by replacing the stub by its Thevenin equivalent. Clearly, looking into a stub where a voltage pulse ${}_kV_c^i$ is incident will result in an open circuit voltage which, as explained in Section 2.1, is equal to twice the incident voltage. This open-circuit voltage, $2\,{}_kV_c^i$ is then the voltage source that appears in the Thevenin equivalent circuit. Similarly, the input impedance of the stub is simply its characteristic impedance Z_c since, over the duration of a time-step Δt, the flow of currents cannot be influenced by the stub termination. Introducing these modifications results in the circuit shown in Fig. 3.3c, representing conditions at time-step $k\Delta t$. The current is then equal to

$${}_kI = \frac{{}_kV_s - 2\,{}_kV_c^i}{R + Z_c} \tag{3.5}$$

and the voltage across the capacitor is

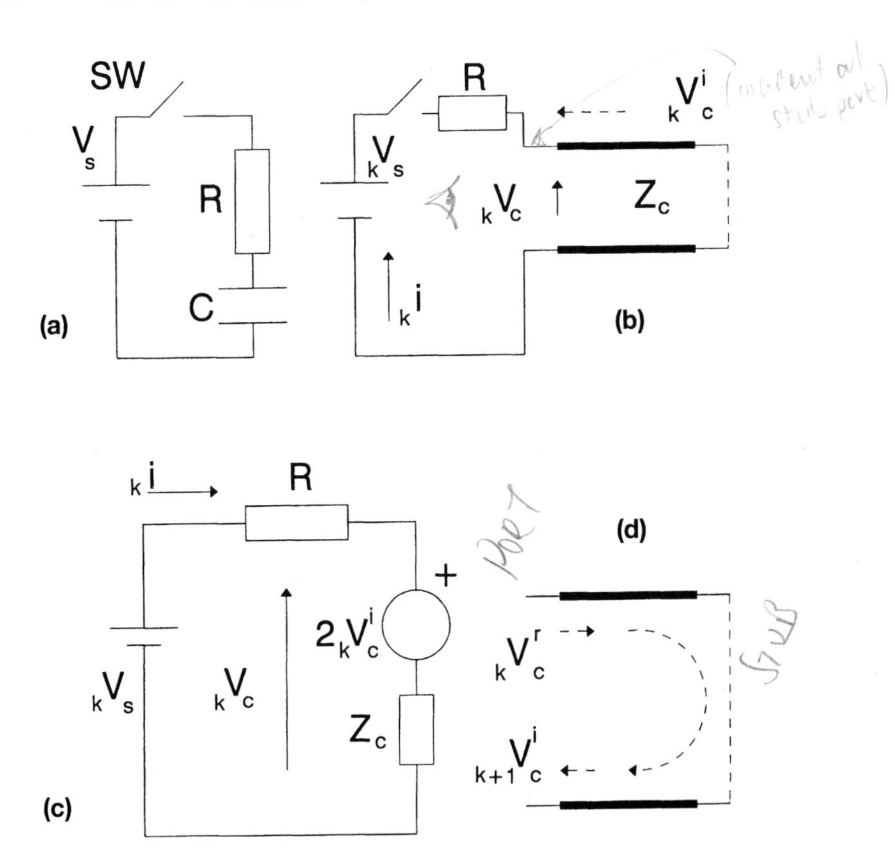

Fig. 3.3 (a) An RC circuit, (b) its TLM model, (c) and Thevenin equivalent. Conditions in the stub are shown in (d).

$$
{}_kV_c = 2\,{}_kV_c^i + {}_kI\,Z_c \tag{3.6}
$$

These expressions show that if the incident voltage ${}_kV_c^i$ is known at time-step k, the voltages and current in the circuit may be calculated. It remains to show how the calculation may advance to the next time-step, $k+1$. To do this, it is necessary to examine what happens in the stub following the incidence of pulse ${}_kV_c^i$ on the port. Clearly, a discontinuity is encountered, and a reflected pulse ${}_kV_c^r$ will be produced and will travel toward the open-circuit termination. The magnitude of the reflected pulse must be such that, when added to the incident pulse, it produces a voltage that is the same as ${}_kV_c$; i.e.,

$$
{}_kV_c = {}_kV_c^i + {}_kV_c^r
$$

The reflected pulse is therefore

$$ {}_kV_c^r = {}_kV_c - {}_kV_c^i \tag{3.7} $$

This pulse is shown schematically traveling toward the termination in Fig. 3.3d. It encounters the open circuit after time $\Delta t/2$, is reflected without change in magnitude or polarity, and travels toward the left to arrive at the port after a further delay of $\Delta t/2$; i.e., at the next time-step, $k + 1$. This pulse is clearly the incident pulse at $k + 1$,

$$ {}_{k+1}V_c^i = {}_kV_c^r = {}_kV_c - {}_kV_c^i \tag{3.8} $$

The new incident voltage may be inserted into Equation (3.5), which is now evaluated at $k + 1$ to obtain ${}_{k+1}I$. Similarly, Equation (3.6) is used to obtain ${}_{k+1}V_c$, and then Equation (3.8) is used to find the incident voltage at time-step $k + 2$. The process is thus repeated for as long as desired.

A simple numerical example is given below to illustrate the process. For the circuit shown in Fig. 3.3, the following values are chosen

$$ V_s = 10 \text{ V} $$

$$ R = 1\ \Omega $$

$$ C = 1 \text{ F} $$

to make calculation by hand easy. Choosing $\Delta t = 0.1$ s gives $Z_C = \Delta t/2C = 0.05\ \Omega$ Following closure of the switch, SW ($k = 0$), strict enforcement of the initial conditions requires that ${}_0i = V_s/R = 10/1 = 10$ A, and that ${}_0V_c = 0$ V. Hence,

$$ {}_0V_c^i = -\frac{{}_0IZ_c}{2} = -\frac{10 \cdot 0.05}{2} = -0.25 \text{ V} $$

The reflected voltage at $k = 0$ is

$$ {}_0V_c^r = {}_0V_c - {}_0V_c^i = 0 - (-0.25) = 0.25 \text{ V} $$

Hence, the incident voltage at $k = 1$ is

$$ {}_1V_c^i = 0.25 \text{ V} $$

Using equation (3.5),

$$ {}_1I = \frac{10 - 2 \cdot 0.25}{1 + 0.05} = 9.0476 \text{ A} $$

The analytical value is obtained from

$$i(t) = \frac{V}{R} e^{-t/(RC)}$$

and at $t = 0.1$ is $i(0.1) = 9.0483$ A.

The voltage across the capacitor, from Equation (3.6), is

$$_1V_c = 2 \cdot 0.25 + 9.0476 \cdot 0.05 = 0.9524 \text{ V}$$

and from Equation (3.7),

$$_1V_c^r = 0.7024 \text{ V}$$

Therefore, $_2V_c^i = 0.7024$ V and $_2I = 8.1859$ A. The analytical value is 8.1873A. Results for the first few time-steps are shown in Table 3.1.

Table 3.1 Computation of circuit shown in Fig. 3.3 ($\Delta t = 0.1$ s)

Time-step k	$_kV_s$	$_kV_c^i$	$_kI$	$_kV_c$	$_kV_c^r$	*Analytical result i(t)*
0	10	–0.25	10	0	0.25	10
1	10	0.25	9.0476	0.9524	0.7024	0.0483
2	10	0.7024	8.1859	1.8141	1.1117	8.1873
3	10	1.1117	7.4063	2.5937	1.482	7.4082
4	10	1.482	6.7009	3.299	1.817	6.7032
5	10	1.817	6.0628	3.9372	2.1202	6.0653

The time-constant of this circuit is $RC = 1$ s, and the time-step chosen for Table 3.1 is $\Delta t = 0.1$ s. To illustrate the stability of the TLM method, results are also shown in Table 3.2 for the case when $\Delta t = 1$ s.

Table 3.2 Computation of circuit shown in Fig. 3.3 ($\Delta t = 1$ s)

Time-step k	$_kV_s$	$_kV_c^i$	$_kI$	$_kV_c$	$_kV_c^r$	*Analytical result i(t)*
0	10	–2.5	10	0	0.25	10
1	10	2.5	3.3333	6.6667	4.1667	3.6787
2	10	4.1667	1.1111	8.8889	4.7222	1.353
3	10	4.7222	0.3704	9.6296	4.9074	0.4978
4	10	4.9074	0.1234	9.8765	4.9691	0.183
5	10	4.9691	$4.115\ 10^{-2}$	9.9588	4.9897	$6.74\ 10^{-3}$

Using such a large time-step has resulted in a deterioration in accuracy, but the method remains stable.

An alternative approach is to use a simpler initial condition that does not strictly enforce conditions at $t = 0$. This can be done by choosing ${}_0V_c^i = 0$ V rather than the correct value of –0.25 V used in Table 3.1. Using this approximate initial condition gives the results shown in Table 3.3.

Table 3.3 Computation of circuit shown in Fig. 3.3 ($\Delta t = 0.1$ s) using approximate initial conditions

Time-step k	${}_kV_s$	${}_kV_c^i$	${}_kI$	${}_kV_c$	${}_kV_c^r$	*Analytical result i(t)*
0	10					10
1	10	0	9.5238	0.4762	0.4762	9.0483
2	10	0.4762	8.6168	1.3832	0.9070	8.1873
3	10	0.9070	7.7961	2.2039	1.2968	7.4082
4	10	1.2968	7.0536	2.9463	1.6495	6.7032
5	10	1.6495	6.3819	3.6181	1.9686	6.0653

The current during the first time-step is effectively the average value of the current calculated analytically at $t = 0$ and $t = 0.1$ s. When small time-steps are used, the errors associated with the approximate initial condition may be acceptable.

The discrete TLM stub model of a capacitor can be related to the associated discrete model described in Ref. [1].

A discrete model associated with the trapezoidal rule of integration is derived as follows. The voltage and current in a capacitor are related by

$$i(t) = C\frac{dV(t)}{dt}$$

Using the trapezoidal rule this expression may be written as

$$\frac{{}_{k+1}I + {}_kI}{2} = C\frac{{}_{k+1}V - {}_kV}{\Delta t}$$

where the subscript k indicates a quantity at time $k\,\Delta t$. Rearranging this equation gives

$${}_{k+1}I = \frac{{}_{k+1}V}{(\Delta t/2C)} - \left[\frac{{}_kV}{(\Delta t/2C)} + {}_kI\right] \tag{3.9}$$

A discrete circuit model associated with this equation is shown in Fig. 3.4.

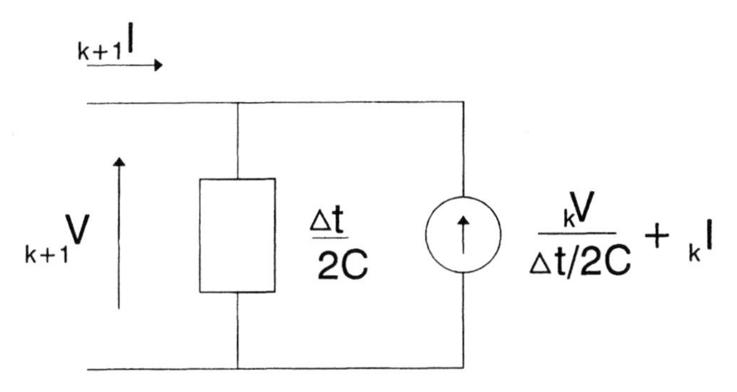

Fig. 3.4 Discrete equivalent model of a capacitor

The TLM algorithm is equivalent to that described by Equation (3.9). Using the notation of Fig. 3.2, where ${}_kV^i$, ${}_kV^r$, ${}_kI^i$, and ${}_kI^r$ are the incident and reflected voltages and currents at the stub port, gives

$$ {}_kV = {}_kV^r + {}_kV^i \tag{3.10} $$

$$ {}_kI = \frac{{}_kV^r - {}_kV^i}{Z_c} \tag{3.11} $$

also,

$$ {}_{k+1}I = \frac{{}_{k+1}V^r - {}_{k+1}V^i}{Z_c} = \frac{{}_{k+1}V^r + {}_{k+1}V^i - 2{}_{k+1}V^i}{Z_c} $$

The sum of the first two terms in the brackets is equal to ${}_{k+1}V$, therefore

$$ {}_{k+1}I = \frac{{}_{k+1}V - 2{}_{k+1}V^i}{Z_c} = \frac{{}_{k+1}V}{Z_c} - \frac{2{}_{k+1}V^r}{Z_c} \tag{3.12} $$

From Equations (3.10) and (3.11),

$$ {}_kV + Z_c \cdot {}_kI = 2{}_kV^r $$

and substituting into Equation (3.12) gives

$$ {}_{k+1}I = \frac{{}_{k+1}V}{Z_c} = \left(\frac{{}_kV}{Z_c} - {}_kI\right) \tag{3.13} $$

Since $Z_c = \Delta t/(2C)$, Equation (3.13) is identical to (3.9).

3.2 "LINK" AND "STUB" MODELS OF INDUCTORS

Consider inductor L shown in Fig. 3.5a. A model based on a link line is shown in Fig. 3.1b. Let us now examine what the parameters of such a model should be. Assuming that the inductance and capacitance per unit length of the link line are L_d and C_d, respectively, and that its length and transit time are $\Delta\ell$ and Δt, then L_d must be chosen so that

$$L_d \Delta\ell = L$$

The velocity of propagation on the line is

$$u = \frac{\Delta\ell}{\Delta t} = \frac{1}{\sqrt{L_d C_d}}$$

Therefore,

$$C_d = \left(\frac{\Delta\ell}{\Delta t}\right)^2 \frac{1}{L_d}$$

and the line characteristic impedance is

$$Z_L = \sqrt{\frac{L_d}{C_d}} = \frac{L}{\Delta t} \tag{3.14}$$

A line with characteristic impedance Z_L behaves as an inductor L, but with an associated capacitance equal to

$$C_e = C_d \cdot \Delta\ell = \frac{(\Delta t)^2}{L} \tag{3.15}$$

C_e may be regarded as a modeling error but, as discussed in the previous section, with a suitable time-step choice, it may be regarded as the stray capacitance associated with any real inductor.

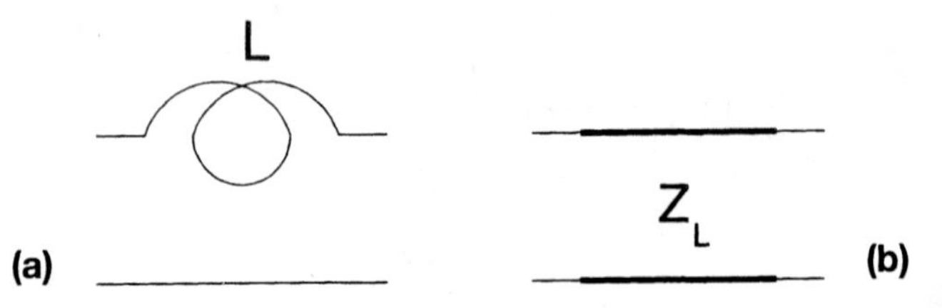

Fig. 3.5 (a) an inductor and (b) its link model

As in the case of a capacitor, an alternative "stub" model may also be developed, as shown in Fig. 3.6. The stub model representing the inductor is terminated by a short circuit because, to emphasize inductive behavior, current and, hence, storage in the magnetic field must be maximized. The round-trip time for the stub is chosen to be Δt.

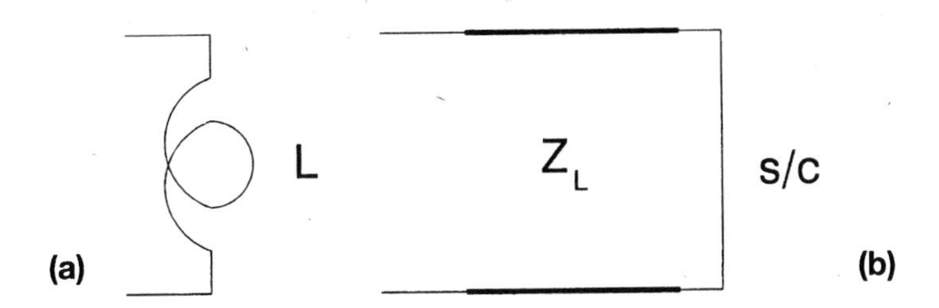

Fig. 3.6 (a) an inductor and (b) its stub model

Using the same procedure as for the capacitance stub, L_d is chosen so that

$$L_d \Delta \ell = L$$

The propagation velocity is

$$u = \frac{\Delta \ell}{(\Delta t/2)} = \frac{1}{\sqrt{L_d C_d}}$$

Hence,

$$C_d = \frac{(\Delta t)^2}{4L\,\Delta \ell}$$

The characteristic impedance of the inductive stub is

$$Z_L = \sqrt{\frac{L_d}{C_d}} = \frac{2L}{\Delta t} \tag{3.16}$$

The associated error or parasitic capacitance is

$$C_e = C_d\,\Delta \ell = \frac{(\Delta t)^2}{4L}$$

Alternatively, the stub model of an inductor may be examined in the frequency domain. Using Equation (2.10) for the case when the load impedance is zero (short-circuit termination) gives for the input impedance

$$Z_{in} = j\, Z_L \tan\left(\frac{\omega \Delta t}{2}\right)$$

$$= j\frac{2L}{\Delta t}\left[\frac{\omega \Delta t}{2} + \frac{1}{3}\left(\frac{\omega \Delta t}{2}\right)^3 + \ldots\right]$$

$$= j\omega L + j\frac{\omega^3 L (\Delta t)^2}{12} + \ldots$$

As in the case of the capacitive stub, the error in the stub impedance increases as the square of the time-step. The frequency dependence of the error term means that errors also increase as the frequency is increased.

Let us now look at the simple inductive circuit shown in Fig. 3.7a. The TLM model of this circuit is shown in Fig. 3.7b, where the inductor has been replaced by a stub equivalent. Replacing this stub by its Thevenin equivalent results in the circuit shown in Fig. 3.7c.

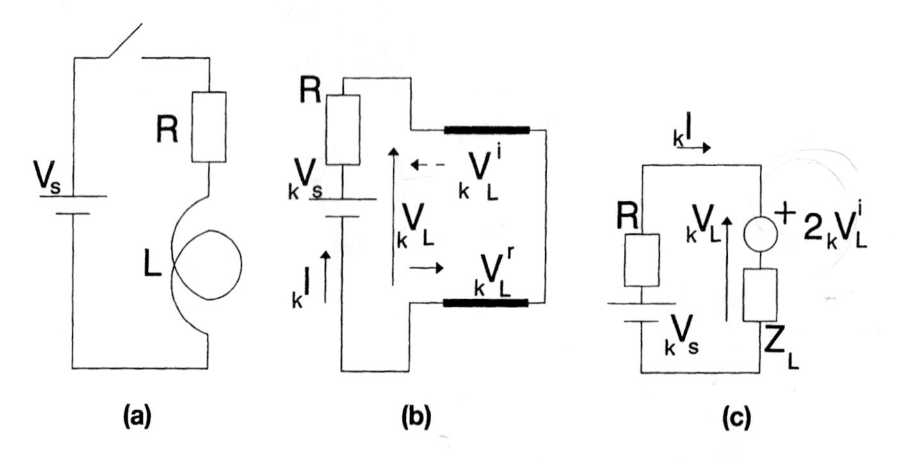

Fig. 3.7 (a) an LR circuit, (b) its TLM model, and (c) its Thevenin equivalent

The current at time $k\Delta t$ is

$$_kI = \frac{_kV_s - 2\,_kV_L^i}{R + Z_L}$$

The voltage across the inductor is

$$_kV_L = 2\,_kV_L^i + {_kI}\, Z_L$$

and the reflected voltage may be obtained from

$${}_kV_L^r = {}_kV_L - {}_kV_L^i$$

The new incident voltage is the voltage pulse that results from the reflection of ${}_kV_L^r$ from the short circuit termination; i.e.,

$${}_{k+1}V_L^i = -{}_kV_L^r = {}_kV_L^i - {}_kV_L \tag{3.17}$$

The only substantial difference in the algorithm for the capacitive and inductive circuits is in Equations (3.8) and (3.17) and results from the different stub termination for the capacitor (open circuit) and for the inductor (short circuit).

The stub TLM model of an inductor results in an algorithm corresponding to the trapezoidal rule of integration. Following a similar procedure to that adopted for a capacitor,

$$V(t) = L\frac{di(t)}{dt}$$

or

$$\frac{{}_{k+1}V + {}_kV}{2} = L\,\frac{{}_{k+1}I - {}_kI}{\Delta t}$$

Rearranging this expression gives

$${}_{k+1}V = {}_{k+1}I\,\frac{2L}{\Delta t} - \left({}_kV + \frac{2L}{\Delta t}\,{}_kI\right) \tag{3.18}$$

A discrete circuit model associated with Equation (3.18) is shown in Fig. 3.8.

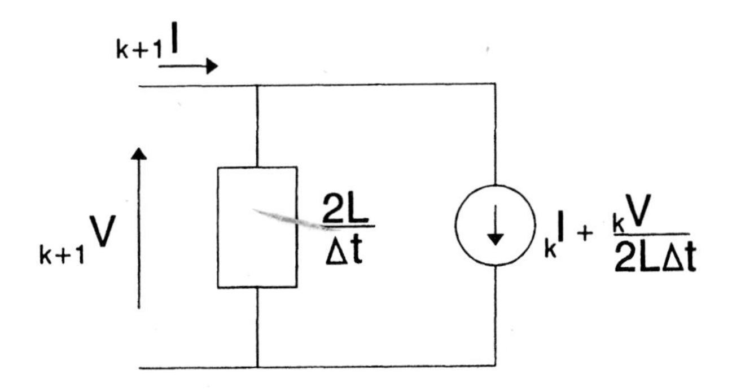

Fig. 3.8 Discrete equivalent model of an inductor

From the TLM stub equivalent of the inductor,

$$ {}_kV = {}_kV^r + {}_kV^i $$

$$ Z_L\, {}_kI = {}_kV^r - {}_kV^i $$

Adding these two expressions gives

$$ {}_kV + Z_L\, {}_kI = 2\,{}_kV^r $$

The current at time-step $k+1$ is

$$ {}_{k+1}I = \frac{{}_{k+1}V^r - {}_{k+1}V^i}{Z_L} = \frac{{}_{k+1}V - 2\,{}_{k+1}V^i}{Z_L} = \frac{{}_{k+1}V + 2\,{}_kV^r}{Z_L} $$

Substituting for ${}_kV^r$ and solving for ${}_{k+1}V$ gives

$$ {}_{k+1}V = {}_{k+1}I\, Z_L - \left({}_kV + Z_L\, {}_{k+1}V^i \right) \tag{3.19} $$

Since $Z_L = 2L/\Delta t$, the algorithm described by Equation (3.19) is identical to that of Equation (3.18). Further details of link and stub models for inductors and capacitors may be found in Refs. [2–4].

3.3 EXAMPLES OF MIXED LINK AND STUB MODELS

In the last two sections, two types of modeling—termed link and stub—were described. It is instructive to consider now their general properties and how they may be used best in modeling applications.

First, both types of modeling are unconditionally stable. This is easily confirmed by the nature of the network, which consists of positive, passive components and is thus stable at all frequencies.

Second, errors are introduced during the modeling process (the solution method being exact). Errors may be expressed in terms of parasitic components; e.g., parasitic inductance in the model of a capacitor. The applied scientist and engineer is thus in a position to ascertain almost intuitively the significance of these errors in the expected response. Moreover, it could be argued that this modeling error, which in physical terms appears as a parasitic component, may be regarded as an additional stray component associated with the real component being modeled. Thus, with proper choice of time-step and control of the modeling error, the description of the component in the model may be closer to reality than an ideal component.

Third, the choice of stub or link modeling results in different algorithm types. Stub modeling results in an implicit solution routine in that, for a network modeled by stubs, a system of simultaneous equations is formed where the system matrix involves the entire network. For large networks, this type of modeling may lead to inefficient solution techniques involving the inversion of very large matrices. In contrast, in a network modeled with links, the solution routine is explicit. The complexity of the equations obtained is independent of the number of nodes or components in the network. This approach is advantageous for large, complex networks.

In general, experienced modelers tend to use a mixture of link and stub modeling according to circumstances. The best choice depends on network complexity and the minimization of modeling errors. It should come as no surprise that the same system may be modeled in several different ways. An example of the options available to the modeler is given for the circuit shown in Fig. 3.9a. An all-link model of this circuit is shown in Fig 3.9b. Impedances Z_1, Z_2, and Z_c are determined as described in Sections 3.1 and 3.2. If inductor L_2 is very small, the associated error capacitance can be very large, as shown by Equation (3.15). This parasitic capacitance will appear in parallel with C and may, in fact, be a significant fraction of it. In such cases, either Δt must be reduced to very small

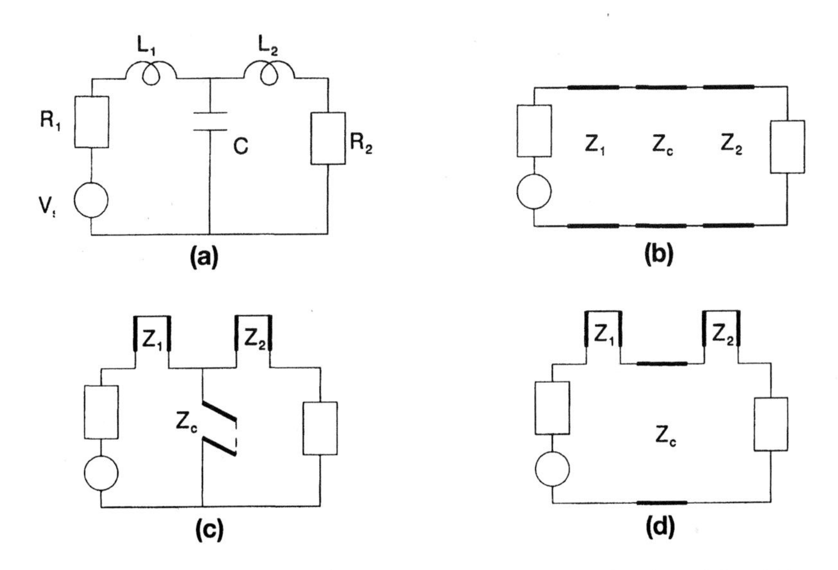

Fig. 3.9 (a) lumped component circuit, (b) all-link model, (c) all-stub model, and (d) mixed model

values or, alternatively, the error capacitance C_e associated with L_2 must be subtracted from C and Z_c chosen to model C–C_e. The modeler thus has various options available to construct the most appropriate model. The solution for the network shown in Fig. 3.9b proceeds by writing four decoupled equations for scattering at the four nodes shown.

An alternative approach is to use an all-stub model as shown in Fig. 3.9c. Solution proceeds by replacing each stub with its Thevenin equivalent circuit and solving the resulting coupled equations for the three unknown incident voltages.

Finally, a mixed stub and link model may be employed as shown in Fig. 3.9d. In solving this network, two decoupled equations are formed for the segments to the left and to the right of the link line. In this configuration, the error capacitance associated with L_2 does not appear in parallel to C, and the model is inherently more accurate compared to the all-link model. Thus, a judicious choice of model can minimize errors and computational effort. The reader could attempt solution of this network using the different models suggested and compare results with other more conventional techniques. A choice of $L_2 << L_1$ results in a stiff network (two very different time constants). Application of the TLM models described here to such a network results in stable solutions irrespective of the time-step chosen.

3.4 MODELING OF NONLINEAR ELEMENTS

All physical systems and components exhibit nonlinear behavior under certain conditions and, in many cases, the operating range and failure modes are determined by nonlinearities. An understanding of nonlinear behavior is therefore fundamental to the design of systems. There are no generally applicable analytical tools that may be applied to the solutions of nonlinear problems. Therefore, in practical situations, it is necessary to resort to numerical techniques to solve such problems.

In this section, the treatment of nonlinearities in TLM models is described.

3.4.1 Nonlinear Resistor

The manner in which nonlinear resistors may be included in a TLM model was first described in Ref. [4]. It is generally desirable to isolate, as far as possible, nonlinear components from the solution algorithm for large networks. This allows for a more efficient treatment and faster solution convergence. The approach adopted is best described by the example of the circuit shown in Fig. 3.10a. It is assumed that the nonlinear resistor R_n is described by a voltage-current relationship of the form $i = A\, v^{\alpha}$. The capacitor may be modeled by a stub to obtain the circuit shown in Fig. 3.10b. Replacing the circuit to the right of A–A′ by its Thevenin equivalent results in the circuit shown in Fig. 3.10c. It should be noted that this last circuit may be solved at each time-step without any direct reference to the nonlinearity. This is particularly important if the circuit to the left of A–A′ is a large and complicated one. The nonlinearity does not appear in the solution routine for this circuit. In particular, the impedance matrix contains Z_C and other linear components (such as R) and thus need be inverted only once and used repeatedly to obtain the solution. The nonlinearity influences the solution during the calculation of the incident voltage at the next time-step. The solution procedure is as follows.

From the circuit shown in Fig. 3.10c, the voltage across the capacitor at time-step k is

$$_kV_C = \frac{Z_C}{R + Z_C}\left({_kV_S} - 2\,{_kV^i} \right) \tag{3.20}$$

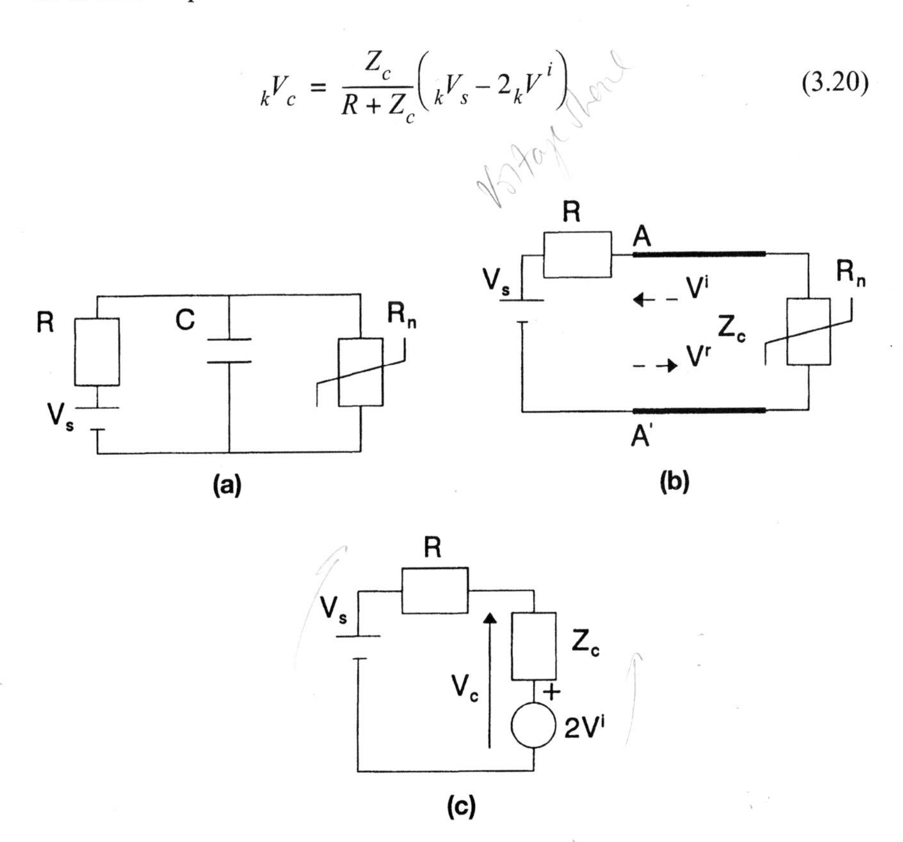

Fig. 3.10 (a) a nonlinear circuit, (b) its TLM model, and (c) the Thevenin equivalent

Hence, the reflected voltage is

$$ {}_kV^r = {}_kV_c - {}_kV^i \tag{3.21} $$

After time $\Delta t/2$, the voltage incident on the nonlinear load will be ${}_kV^r$, and the reflected voltage there will be what after another delay $\Delta t/2$ becomes the new incident voltage at A–A′; namely, ${}_{k+1}V^i$. Hence, the voltage and current at the nonlinear load are

$$ V_L = {}_kV^r + {}_{k+1}V^i $$

$$ I_L = \frac{{}_kV^r - {}_{k+1}V^i}{Z_c} $$

Substituting these expressions into the voltage-current relation describing the load gives

$$ \frac{{}_kV^r - {}_{k+1}V^i}{Z_c} = A\left({}_kV^r + {}_{k+1}V^i\right)^{\alpha} \tag{3.22} $$

Since ${}_kV^r$ is known from Equation (3.21), Equation (3.22) may be solved by iterative methods to obtain ${}_{k+1}V^i$.

This value is substituted into Equation (3.20) to obtain ${}_{k+1}V_c$, and the process is repeated for as long as desired. The attractive feature of this approach is that the nonlinearity is dealt with entirely in Equation (3.22), which is independent of the complexity of the circuit to the left of A–A′. Similarly, the main solution algorithm [which, in this simple case, is represented by (3.20), but for a complex network may involve large matrix inversion] is entirely independent of the nonlinearity. In practice, even if a convenient component, such as capacitor C, is not present to decouple the nonlinearity from the rest of the circuit, sufficient stray capacitance or inductance may be present to construct a suitable model.

3.4.2 The Switching Element

A common circuit component is the electrical switch. This may be of electromechanical or electronic design (e.g., a thyristor). From the modeling point of view, the switch may be viewed as a strongly nonlinear component and, if the voltage-current characteristic is known, the techniques of the previous section may be used. In many cases, however, such a characteristic either is not available, or it is not considered necessary to describe switching behavior in such detail. The conventional approach to switch modeling is then to represent the switch by a resistor with zero value when it is closed and a very large value when it is open. The disadvantage of this technique is that the network impedance matrix has to be recalculated each time any switch changes state. In networks, with several switches changing state frequently, as is commonly the case in power electronics, many matrix calculations are necessary. An alternative TLM-based approach is possible that results in a single, constant network matrix, irrespective of the state of switches in the network [5, 6]. In implementing this method, a switch is modeled as a small capacitance when open and a small inductance when closed. To model the switch shown in Fig. 3.11a, a stub of characteristic impedance Z_{SW} is selected, as shown in Fig. 3.11b, and terminated by an ideal switch S. When S is closed, the stub parameters seen across A–B are those of a small inductance (short circuit stub) and, thus, the stub approximately represents the closed switch, SW. Similarly, when S is open, the open circuit stub represents a small capacitance and, thus, the situation when SW is open. The value of Z_{SW} must be chosen with care, taking account of the desired time-step and switch parameters. As far as the rest of the circuit is concerned, the switch (branch A–B) is replaced by its Thevenin equivalent circuit as shown in Fig. 3.11c. The impedance of this equivalent, Z_{SW}, retains a constant value irrespective of whether SW is open or closed. Thus, the formulation and solution of circuit equations is particularly simple. The condition of the switch is taken into account when the value of the incident voltage at the next time-step is calculated; i.e.,

$$ {}_{k+1}V^{i} = \pm {}_{k}V^{r} \tag{3.23} $$

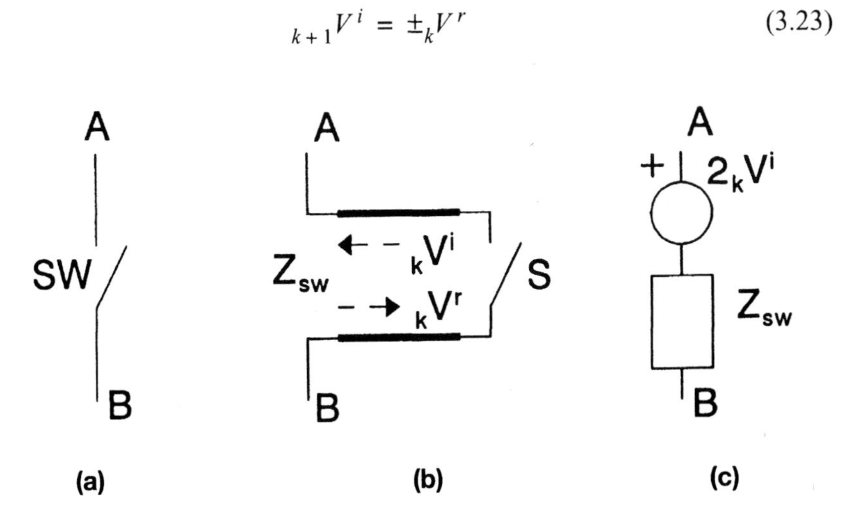

Fig. 3.11 (a) a switching element, (b) its TLM model, and (c) the Thevenin equivalent

where the plus sign is selected when the switch is open, and the minus when the switch is closed. Further details of the application of this method may be found in the references given.

3.4.3 Nonlinear Reactive Components

In many applications, it is necessary to account for the nonlinear behavior of inductors and capacitors. Examples are saturation and hysteresis for inductors and corona effects on the capacitance of high-voltage lines. A special feature of reactive components is that they store energy, so any nonlinear solution algorithm must be consistent with the usual conservation laws. Attempts to deal with such problems in TLM have been reported in Refs. [7] and [8], with varying degrees of success. The most general TLM formulation is presented in Refs. [9] and [10] and is briefly described here.

The voltage drop across a current-dependent inductor $L(i) = d\lambda/di$ is

$$V_L = \frac{d\lambda}{dt} = L(i)\frac{di}{dt} = L(i)\left[1\,\frac{di}{dt}\right] = L(i)\,V_{Lu}$$

where the expression in brackets represents voltage drop V_{Lu} across a 1 H inductor in which current i flows. A stub model of the unit inductor may be constructed in the usual way to give $V_{lu} = Z_{Lu}i + 2V^i_{Lu}$, where $Z_{Lu} = 2/\Delta t$. Thus, the voltage across the inductor $L(i)$ is

$$V_L = L(i)\left[Z_{Lu}i + 2V^i_{Lu}\right] \tag{3.24}$$

Equation (3.24) may be used with the circuit equations to find the current. At each time-step, the value of $L(i)$ is updated, and V_{Lu} is calculated. The new incident voltage is then

$$_{k+1}V^i_{Lu} = {}_kV^i_{Lu} - {}_kV_{Lu} \tag{3.25}$$

A similar approach may be adopted to deal with a voltage-dependent capacitance, $C(v)$. The voltage across such a capacitor is

$$V_c = \frac{2}{C(v)} = \frac{1\,V_{cu}}{C(v)}$$

where V_{cu} is the voltage across a 1 F capacitor; i.e.,

$$V_{cu} = Z_{cu}i + 2V^i_{cu} \tag{3.26}$$

where $Z_{cu} = \Delta t/2$.

The solution procedure is similar to that for a nonlinear inductor, but with the new incident voltage obtained from

$$_{k+1}V^{i}_{cu} = {}_{k}V_{cu} - {}_{k}V^{i}_{cu} \tag{3.27}$$

Further details and examples of the application of this method to nonlinear problems may be found in Ref. [10].

3.5 MODELING OF COUPLED ELEMENTS

Magnetic coupling between components may also be described using TLM. A typical configuration is shown in Fig. 3.12a. Inductors L_1 and L_2 may be modeled in the normal way by short-circuit stubs. Mutual coupling is modeled by current-controlled voltage sources. The resulting

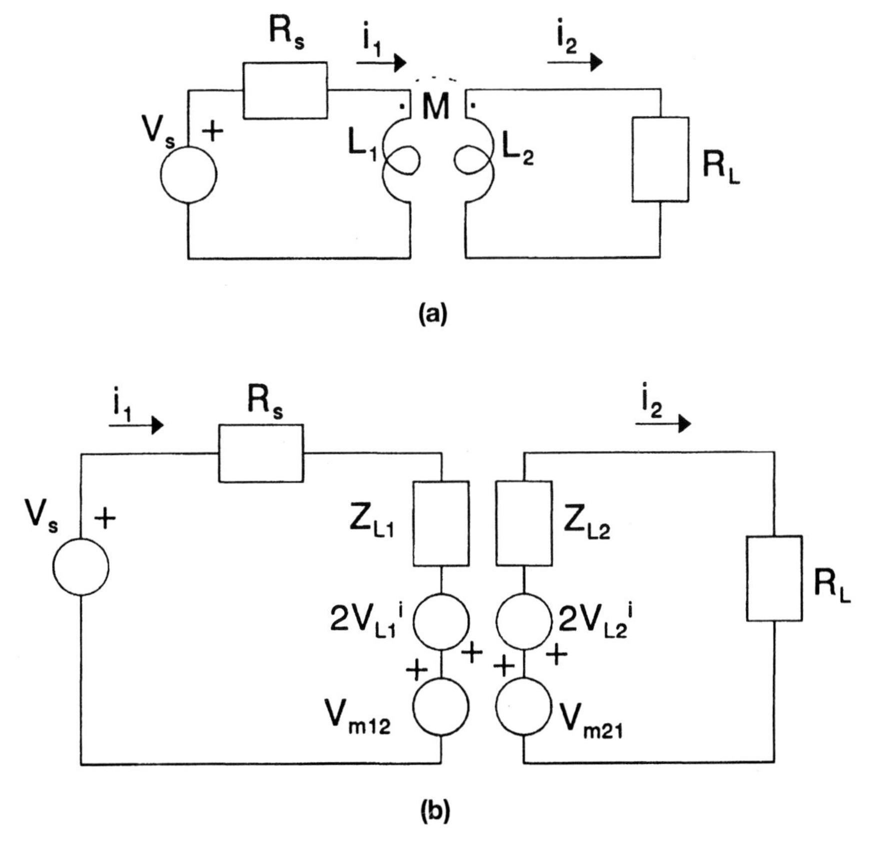

Fig. 3.12 (a) two coupled circuits and (b) the Thevenin equivalent of their TLM model

Thevenin equivalent is shown in Fig. 3.12b. The controlled sources representing terms of the type Mdi/dt are

$$V_{m12} = Z_m i_2 + 2V^i_{m12}$$

$$V_{m21} = Z_m i_1 + 2V^i_{m21} \tag{3.28}$$

where $Z_m = 2\,M/\Delta t$.

The solution proceeds by writing KVL for the two circuits in Fig. 3.12b.

$$(Z_{L1} + R_s)\,i_1 - Z_m i_2 = V_s - 2\left(V^i_{L1} - V^i_{m12}\right)$$

$$-Z_m i_1 + (Z_{L2} + R_L)\,i_2 = 2\left(V^i_{m21} - V^i_{L2}\right) \tag{3.29}$$

The equations are solved to find i_1 and i_2, and thus the total voltages are

$$V_{L1} = 2V^i_{L1} + i_1\,Z_{L1}$$

$$V_{L2} = 2V^i_{L2} + i_2\,Z_{L2}$$

$$V_{m12} = 2V^i_{m12} + Z_m\,i_2$$

$$V_{m21} = 2V^i_{m21} + Z_m\,i_1 \tag{3.30}$$

The incident voltages at the next time-step are

$${}_{k+1}V^i_{L1} = {}_kV^i_{L1} - {}_kV_{L1}$$

$${}_{k+1}V^i_{L2} = {}_kV^i_{L2} - {}_kV_{L2}$$

$${}_{k+1}V^i_{m12} = {}_kV^i_{m12} - {}_kV_{m12}$$

$${}_{k+1}V^i_{m21} = {}_kV^i_{m21} - {}_kV_{m21} \tag{3.31}$$

These values are substituted into Equation (3.29) to calculate the currents ${}_{k+1}i_1$, ${}_{k+1}i_2$ at the next time-step, and this process is repeated for as long as desired. Further details of this technique and its application may be found in Refs. [5], [11], and [12].

3.6 GENERALIZED DISCRETE TLM MODELING

In the previous sections, various TLM-based models were presented for solving electrical circuits. In this section, this approach is consolidated into a general discrete transform method for solving electrical networks, and it is also extended to encompass the modeling of general physical systems described by a system of integro-differential equations. The TLM discrete transform can be applied as easily as the familiar Laplace Transform, with the added advantage that results are obtained directly in the discrete time-domain, and there is no need for an inverse transform. The unconditional stability of the TLM algorithm ensures that the TLM discrete transform may be applied to the solution of general networks, including stiff networks. Discrete transforms for linear and nonlinear elements are shown in Table 3.4. Further details and application examples may be found in Refs. [13] and [10].

Extension of the transform to deal with other physical systems described by integro-differential equations is based on the observation that derivative terms may be represented in circuit terms by a voltage drop in an inductance. Similar analogies may be established between integral terms and the voltage across a capacitor, proportional terms and the voltage drop across a resistor, and, finally, between coupled differential terms and the voltage drop due to mutual inductance. Let us consider the solution of the following equation:

$$a_0 + a_1(z, t)\frac{dx}{dt} + a_2(w, t)\int x dt + bx = y(t) \qquad (3.32)$$

where a_0, a_1, and a_2 are coefficients that depend on t and parameters z and w. Using the discrete transforms from Table 3.4, terms dx/dt and $\int x dt$ are replaced by

$$\left(Z_{Lu}x + 2V^i_{Lu}\right) \text{ and } \left(Z_{cu}x + 2V^i_{cu}\right)$$

respectively, so that Equation (3.32) reduces to

$$x = \frac{y - a_0 - a_1 2V^i_{Lu} - a_2 2V^i_{cu}}{a_1 Z_{Lu} + a_2 Z_{Lu} + b} \qquad (3.33)$$

Table 3.4 Discrete transforms (Reprinted with permission from "Discrete transform technique for solving nonlinear circuits and equations," S.Y.R. Hui and C. Christopoulos, *IEE Proc-A*, 139, 1992)

	Continuous models	Discrete transforms
Constant coefficients		
R	$V_R = R \cdot i$	$V_R = R \cdot i$
L	$V_L = L\, di/dt$	$Z_L = 2L/T$ $2V_L^i$ $V_L = iZ_L + 2V_L^i$
C	$V_c = (1/C)\int i\, dt$	$Z_c = T/(2C)$ $2V_c^i$ $V_C = iZ_c + 2V_c^i$
M_{12}	$V_{M12} = M_{12}\, di2/dt$	$V_{M12} = i_2 Z_{M12} + 2V_{M12}^i$
Varying coefficients		
r	$V_r = r \cdot i$	$V_r = r \cdot i$
$L(i)$	$V_L = d\lambda/dt = L(i) \cdot (di/dt)$ where $L(i) = d\lambda/di$	$L(i)Z_{Lu}$ where $Z_{Lu} = 2/T$ $L(i)2V_{Lu}^i$ $V_L = L(i)[iZ_{Lu} + 2V_{Lu}^i]$
$C(v)$	$V_c = \int \frac{i}{C(v)} dt$	$\frac{Z_{cu}}{C(v)}$ where $Z_{cu} = T/2$ $\frac{2V_{cu}^i}{C(v)}$ $V_c = [iZ_{cu} + 2V_{cu}^i]/C(v)$
$M_{12}(i_2)$	$V_{M12} = d\lambda_{12}/dt = M_{12}(i_2)\, di_2/dt$ where $M_{12}(i_2) = d\lambda_{12}/di_2$	$V_{M12} = M_{12}(i_2)[i_2 Z_{Lu} + 2V_{Lu}^i]$

Solution then proceeds with the substitution of the incident voltages that represent the initial conditions of the original problem. The value of x is thus calculated, and the total voltages are obtained from

$$ {}_{k}V = Z\,x + 2\,{}_{k}V^{i} $$

The nonlinear coefficients a_1 and a_2 are then updated. The new incident voltages are calculated from expressions of the type

$$ {}_{k+1}V^{i} = \pm\left({}_{k}V - {}_{k}V^{i}\right) $$

depending on the nature of the particular term, and are substituted in Equation (3.33) to obtain the value of x at the next time-step. This procedure is repeated for as long as desired, giving the solution to the equation directly in the time domain. A number of nonlinearities including hysteresis may be dealt with using this method, as explained in Ref. [10]. The TLM discrete transform method has many practical advantages, and its performance is close to that of the third-order Gear method. Further applications, a discussion of errors, and comparisons with other numerical methods may be found in Refs. [10] and [13].

REFERENCES

[1] Chua, L.O., and P.M. Lin. 1975. *Computer-Aided Analysis of Electronic Circuits*. Englewood Cliffs, NJ: Prentice Hall.

[2] Johns, P.B. 1977. Numerical modelling by the TLM method. *Proceedings* of the International Symposium on Large Engineering Systems, ed. A. Wexler. New York: Pergamon Press, 139–151.

[3] Bandler, J.W., et al. 1977. Transmission-line modelling and sensitivity evaluation for lumped network simulation and design in the time-domain. *Journal of the Franklin Institute* 304, no. 1, 15–33.

[4] Johns, P.B., and M. O'Brien. 1980. Use of the transmission line modelling (TLM) method to solve non-linear lumped networks., *Radio and Electronic Engineer* 50, 59–70.

[5] Hui, S.Y.R., and C. Christopoulos. 1989. A discrete approach to the modelling of power electronic switching networks. 20th Annual IEEE Power Electronics Specialists Conference, June 26–29, *Proceedings* Vol 1., 130–137.

[6] Hui, S.Y.R., and C. Christopoulos. 1989. The modelling of networks with frequently changing topology whilst maintaining a constant system matrix. *International Journal of Numerical Modelling* 3, 11–21.

[7] Newcombe, L.A. and J.E. Sitch. 1985. Reactive non-linearities in transmission line models. *Proceedings of IEE-A* 132, 95–98.

[8] Wong, C.C. 1988. Transmission line modelling of non-linear reactive components and mutual inductances. *International Journal of Numerical Modelling* 1, 89–91.

[9] Hui, S.Y.R., and C. Christopoulos. 1991. Non-linear transmission line modelling technique for modelling power electronic circuits. European Power Electronics Conference, Florence. *Proceedings* Vol. 1, 80–84.

[10] Hui, S.Y.R., and C. Christopoulos. 1992. Discrete transform technique for solving non-linear circuits and equations. *Proceedings of IEE-A* 139, 321–328.

[11] Hui, S.Y.R., and C. Christopoulos. 1990. Numerical simulation of power circuits using transmission line modelling. *Proceedings of IEE-A* 137, 379–384.

[12] Hui, S.Y.R., and C. Christopoulos. 1990. A discrete approach to the modelling of power electronic switching networks. *IEEE Transactions on Power Electronics* 5, 398–403.

[13] Hui, S.Y.R., and C. Christopoulos.1991. Discrete transform technique for solving coupled integro-differential equations in digital computers. *Proceedings of IEE-A* 138, 273–280.

4

One-Dimensional TLM Models

The principles of establishing an analogy between electrical circuits and physical phenomena such as wave propagation was discussed in general terms in Section 1.3. This chapter examines more closely the modeling procedure and computation for the one-dimensional case.

4.1 TLM MODEL OF A LOSSY TRANSMISSION LINE

A lossy line is shown in Fig. 4.1, where R and G are the series resistance and shunt admittance, and L and C are the series inductance and shunt capacitance per section of length, Δx.

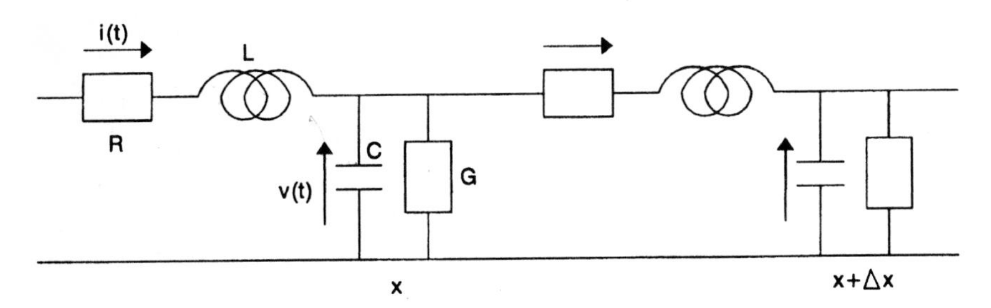

Fig. 4.1 A lossy transmission line

From Kirchhoff's voltage and current laws, the following expressions are obtained:

$$\Delta x \frac{\partial v}{\partial x} = -L\frac{\partial i}{\partial t} - iR \tag{4.1}$$

$$\Delta x \frac{\partial i}{\partial x} = -G\,v - C\,\frac{\partial v}{\partial t} \tag{4.2}$$

where v and i (the voltage and current on the line) are both functions of x and t. These equations are strictly correct in the limit of $\Delta x \to 0$.

They can be manipulated to eliminate the voltage and thus obtain an equation for the current. This can be done by differentiating Equation (4.1) with respect to t, and (4.2) with respect to x, and combining to obtain

$$\frac{\partial^2 i}{\partial x^2} = \frac{GR}{(\Delta x)^2}\, i + \frac{1}{(\Delta x)^2}(GL + RC)\frac{\partial i}{\partial t} + \frac{LC}{(\Delta x)^2}\frac{\partial^2 i}{\partial t^2} \tag{4.3}$$

A similar procedure can be followed to eliminate the current and obtain the following equation for the voltage.

$$\frac{\partial^2 v}{\partial x^2} = \frac{GR}{(\Delta x)^2}\, v + \frac{1}{(\Delta x)^2}(GL + RC)\frac{\partial v}{\partial t} + \frac{LC}{(\Delta x)^2}\frac{\partial^2 v}{\partial t^2} \tag{4.4}$$

Let us now examine the method of computation of the voltage and current on a line consisting of sections such as shown in Fig. 4.2a. The inductance and capacitance may be represented by a transmission line of impedance $Z_o = \sqrt{L/C}$ and of transit time

$$\Delta t = \Delta x/u = \Delta x/\left[1/\sqrt{(L/\Delta x)\,(c/\Delta x)}\right] = \sqrt{LC}$$

Components R and G may be added to this lossless line to produce the TLM equivalent of one line section shown in Fig. 4.2b. Let us also assume that the line consists of ten identical sections and that it is terminated as shown in Fig. 4.2c. It is required to find the voltage and current on the line as a function of x and t when the source voltage $V_s(t)$ is known. The connection points or "nodes" between sections are labeled, starting at $n = 1$ adjacent to the source and extending to $n = 11$ next to the load. Conditions around an arbitrary node n (other than $n = 1$ and $n = 11$, which are somewhat special) and at time-step k are as shown in Fig. 4.3a. At node n, the total voltage ${}_kV_n$ at time-step k will be the result of incident voltages coming from the left (${}_kVL^i_n$) and from the right (${}_kVR^i_n$).

Clearly, the lines to the left and to the right of node n may be replaced for the duration of time-step k by their Thevenin equivalents as shown in Fig. 4.3b. The voltage ${}_kV_n$ can be obtained directly using the parallel generator theorem or Millman's theorem [1] to give:

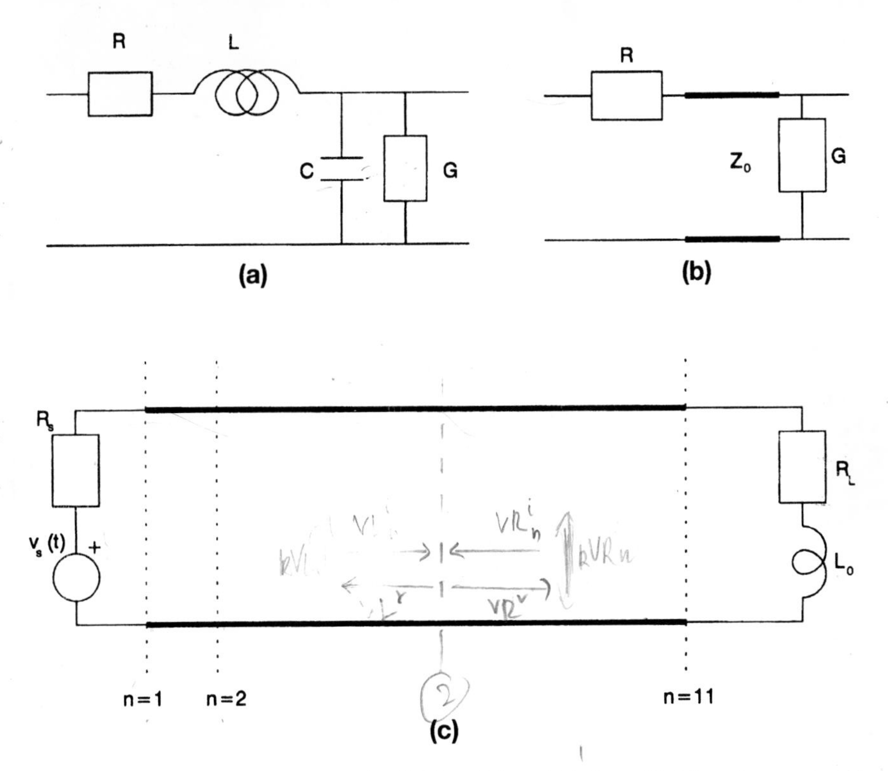

Fig. 4.2 (a) basic line segment, (b) its TLM model, and (c) line with 10 sections, as in (b)

$$ {}_kV_n = \frac{\dfrac{2\,{}_kVL^i_n}{Z_o} + \dfrac{2\,{}_kVR^i_n}{Z_o + R}}{\dfrac{1}{Z_o} + \dfrac{1}{R + Z_o} + G} \tag{4.5} $$

Similarly, the current is

$$ {}_kI_n = \frac{{}_kV_n - 2\,{}_kVR^i_n}{R + Z_o} \tag{4.6} $$

where ${}_kV_n$ is obtained from Equation (4.5).

The total voltage at the input to the line on the left is

$$ {}_kVL_n = {}_kV_n \tag{4.7} $$

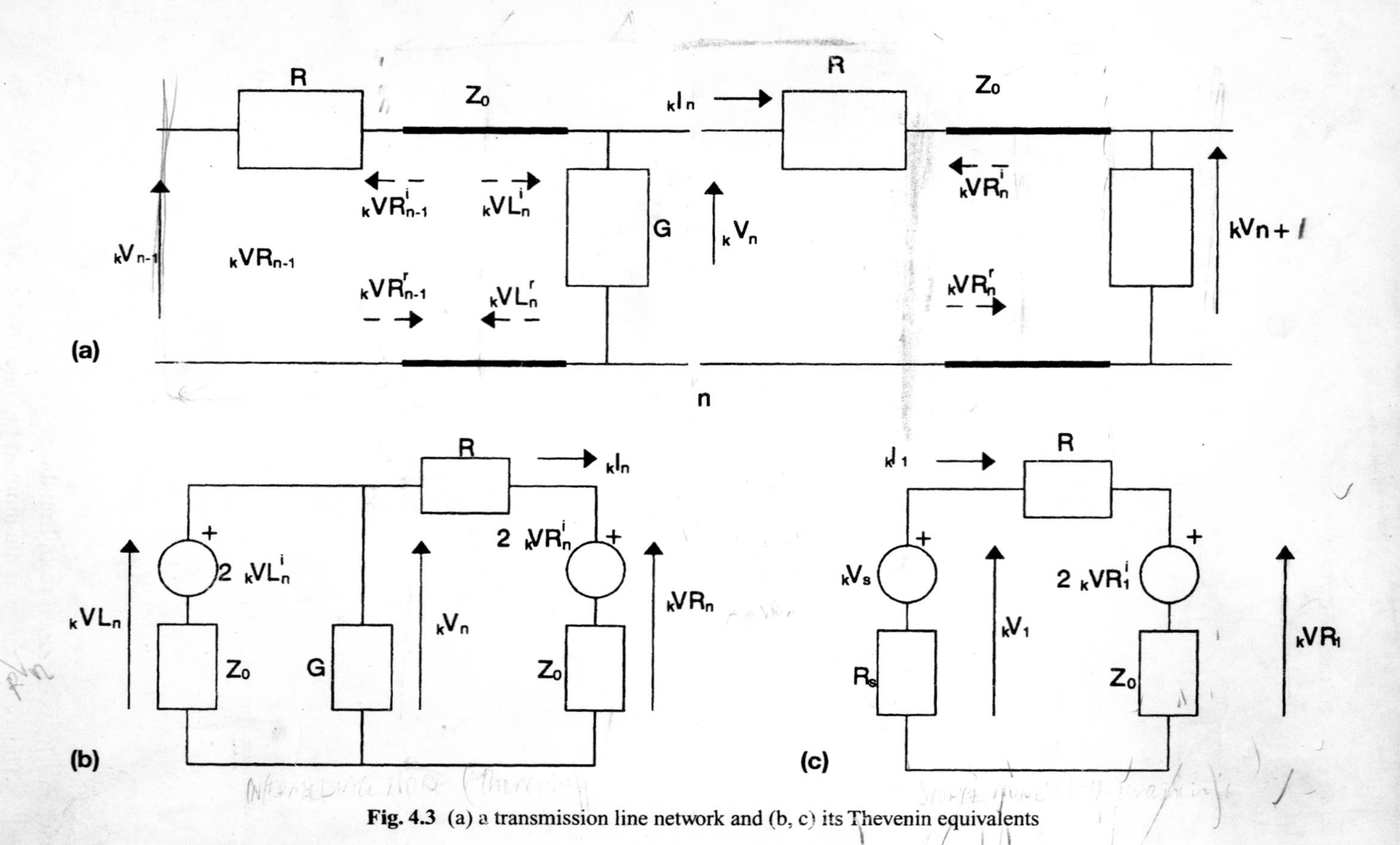

Fig. 4.3 (a) a transmission line network and (b, c) its Thevenin equivalents

and to the line on the right is

$$ {}_kVR_n = 2\,{}_kVR^i_n + {}_kI_n\, Z_o \tag{4.8}$$

where ${}_kI_n$ is obtained from Equation (4.6).

The voltages that are reflected into the lines on the left and on the right are

$$ {}_kVL^r_n = {}_kVL_n - {}_kVL^i_n $$

$$ {}_kVR^r_n = {}_kVR_n - {}_kVR^i_n \tag{4.9}$$

Equations (4.5) through (4.9) express the fact that if the two incident voltages ${}_kVL^i_n$ and ${}_kVR^i_n$ are known, then all other quantities, including the reflected voltages, may be calculated directly. The incident voltages at $t = 0$ are determined from the initial conditions. It remains to show how the incident voltages at time-step $k + 1$ may be obtained from knowledge of conditions at time-step k. This is a straightforward procedure, and it is based on the topology of the network under consideration. The voltage incident from the left on node n at time step $k + 1$, ${}_{k+1}VL^i_n$, is simply what was reflected into the right of node $n - 1$ at the previous time-step k; i.e., ${}_kVR^r_{n-1}$. Similar considerations apply for the incident voltage on node n coming from the right. The new incident voltages are given by

$$ {}_{k+1}VL^i_n = {}_kVR^r_{n-1} $$

$$ {}_{k+1}VR^i_n = {}_kVL^r_{n+1} \tag{4.10}$$

The general principles employed to find conditions at node n, with small modifications, can also be applied to the source ($n = 1$) and load ($n = 11$) nodes.

For the source node, the equivalent circuit is shown in Fig. 4.3c, and it can be readily shown that

$$ {}_kV_1 = \frac{\dfrac{{}_kV_s}{R_s} + \dfrac{2\,{}_kVR^i_1}{R + Z_o}}{\dfrac{1}{R_s} + \dfrac{1}{R + Z_o}} \tag{4.11}$$

$$ {}_kI_1 = \frac{{}_kV_1 - 2\,{}_kVR_1^i}{R + Z_o} \tag{4.12}$$

$$ {}_kVR_1 = 2\,{}_kVR_1^i + {}_kI_1\, Z_o \tag{4.13}$$

$$ {}_kVR_1^r = {}_kVR_1 - {}_kVR_1^i \tag{4.14}$$

$$ {}_{k+1}VR_1^i = {}_kVL_2^r \tag{4.15}$$

For the load node, the circuit is shown in Fig. 4.4a. To proceed further, it is necessary to replace inductor L_o by its TLM discrete model as shown in Fig. 4.4b, where $Z_L = 2L_o/\Delta t$, and the round trip propagation time on the short-circuit stub is Δt. This choice of propagation time for the stub maintains synchronism between pulses in the link lines and the stub. Replacing the link and stub lines by their Thevenin equivalent results in the circuit of Fig. 4.4c. The voltage across the load at time-step k is

$$ {}_kV_{11} = \frac{\dfrac{2\,{}_kVL_{11}^i}{Z_o} + \dfrac{2\,{}_kV^i}{R_L + Z_L}}{\dfrac{1}{Z_o} + \dfrac{1}{R_L + Z_L} + G} \tag{4.16}$$

Similarly, the load current is

$$ {}_kI_L = \frac{{}_kV_{11} - 2\,{}_kV^i}{R_L + Z_L} \tag{4.17}$$

The conditions at the inductor are

$$ {}_kV = 2\,{}_kV^i + {}_kI_L\, Z_L \tag{4.18}$$

$$ {}_kV^r = {}_kV - {}_kV^i \tag{4.19}$$

$$ {}_{k+1}V^i = -{}_kV^r \tag{4.20}$$

The voltage reflected into the link line is

$$ {}_kVL_{11}^r = {}_kV_{11} - {}_kVL_{11}^i \tag{4.21}$$

and the new incident voltage is

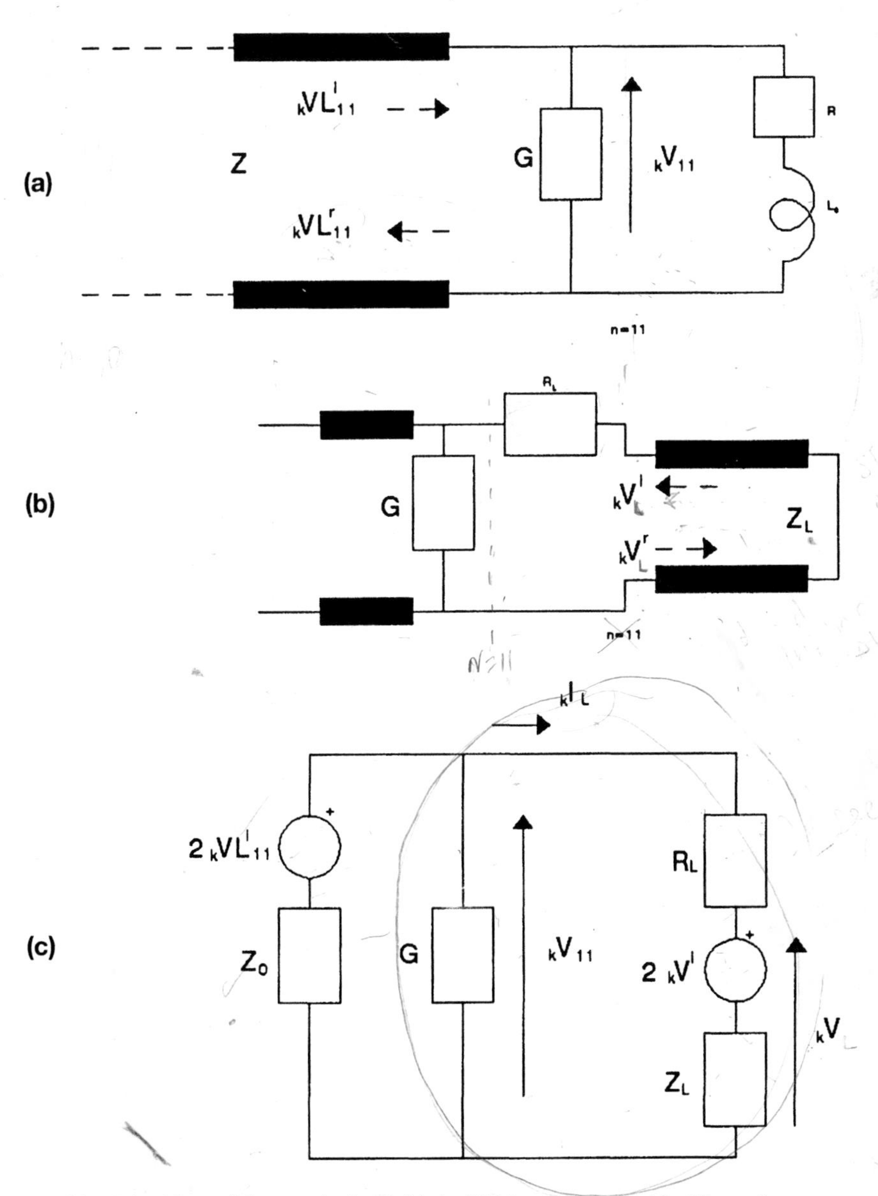

Fig. 4.4 (a) conditions at the load, (b) the TLM model, and (c) its Thevenin equivalent

$$_{k+1}VL^{i}_{11} = {}_kVR^{r}_{10} \tag{4.22}$$

The implementation of the procedure described by Equations (4.5) through (4.22) for the number of desired time-steps provides the solution to the problem shown in Fig. 4.2. A schematic diagram outlining the logical steps followed in this implementation is shown in Fig. 4.5. It consists of the initial problem definition and housekeeping tasks, followed by the "calculation of voltages and currents," "scattering," and "connection." The last three steps are repeated for the desired number of time-steps.

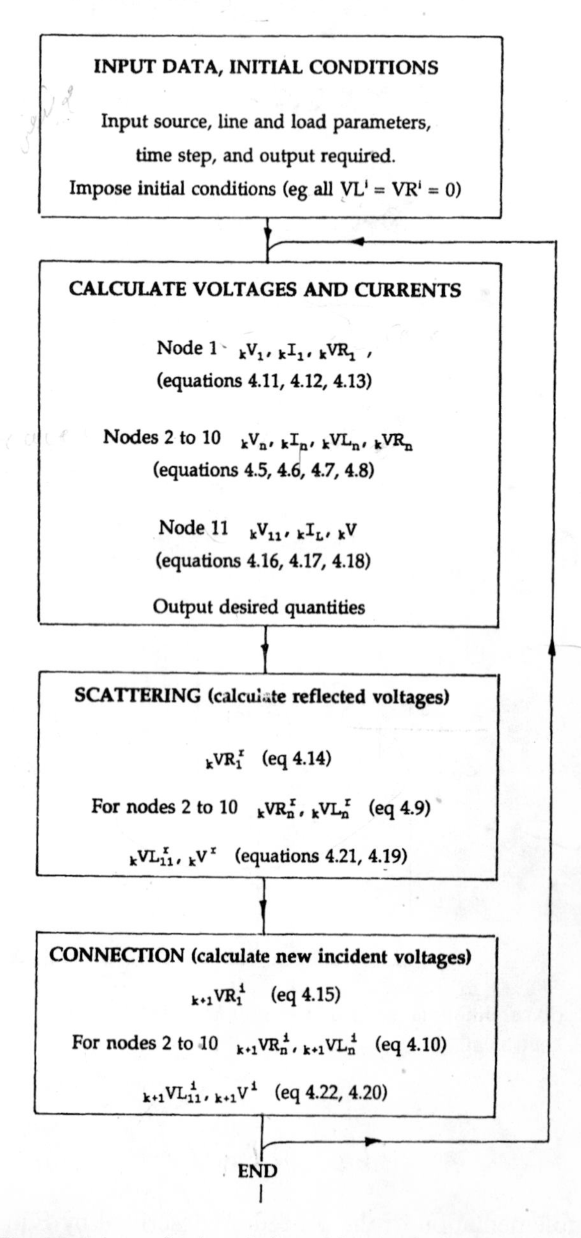

Fig. 4.5 Scheme for the solution of the circuit shown in Fig. 4.2

A practical problem often rises whereby the parameters of the line change with x. An obvious example is of a line formed by joining two lines of a different characteristic impedance. Invariably, this means that the velocity of propagation is also different and, therefore, if the same space length Δx is used throughout, the propagation time will vary along the line. This causes difficulties, because pulses will arrive at different times at nodes of the system. It is then impossible to keep track and combine pulses because they are not synchronized with each other. It is therefore of paramount importance to maintain the same time-step, even on a line with nonuniform parameters. If the problem contains two lines of characteristic impedance and velocity Z_1, u_1, and Z_2, u_2, respectively, then if the same time-step Δt is chosen throughout the problem, the segment length on each line will be different; i.e., $\Delta x_1 = u_1 \Delta t$ and $\Delta x_2 = u_2 \Delta t$. The computation would then proceed as already described, with obvious modifications to account for the change in characteristic impedance. For example, at the junction between two different lines (node n) where the characteristic impedances are Z_1 and Z_2 on the left and on the right, respectively, Equation (4.5) would be modified as follows:

$$
{}_kV_n = \frac{\dfrac{2\,{}_kVL_n^i}{Z_1} + \dfrac{2\,{}_kVR_n^i}{Z_2 + R}}{\dfrac{1}{Z_1} + \dfrac{1}{R + Z_2} + G}
$$

An alternative approach is to maintain the same Δx and Δt throughout the problem, but to take account of changes in parameters by adding extra lumped components. To illustrate this approach, let us consider a case where the line to the right of node n has a capacitance per unit length twice as large as that of the line on the left. In the line to the left of node n, a section of length Δx has inductance and capacitance L and C; hence the time-step and characteristic impedance are:

$$
\Delta t = \sqrt{LC}
$$

$$
Z_1 = \sqrt{L/C}
$$

A section of length Δx to the right of node n will have inductance and capacitance equal to L and $2C$, respectively. This section is shown in Fig. 4.6a. Components L and C may be represented by a transmission line as shown in Fig. 4.6b, where this line has the same parameters as for sections to the left of node n. The required extra capacitance C is represented

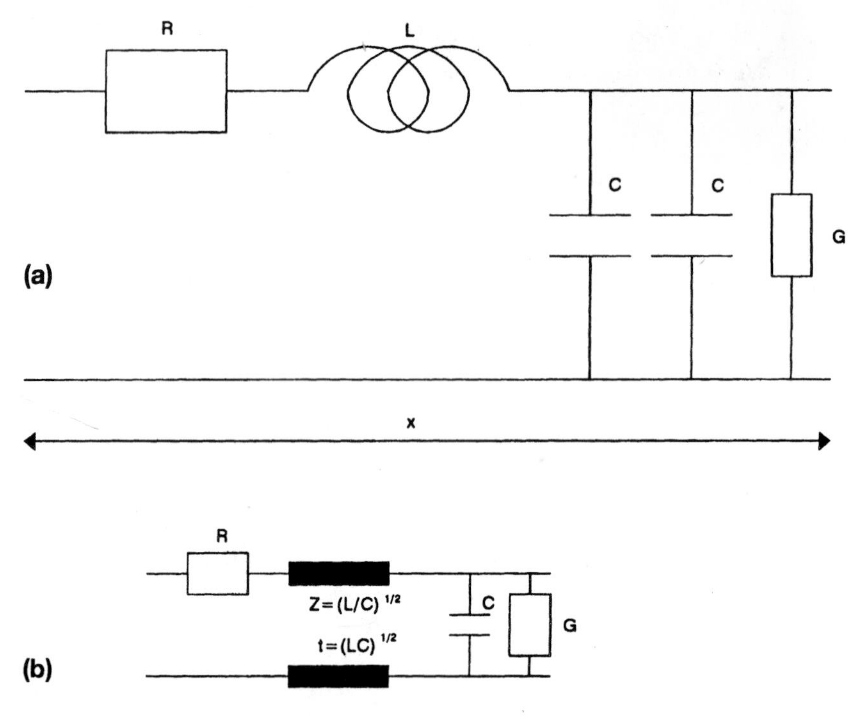

Fig. 4.6 Representation of additional capacitance in (a), as shown in (b), where C is to be represented by a stub.

by a stub of round-trip time Δt and characteristic impedance $Z_C = \Delta t/2C$. In this way, both Δt and Δx remain the same throughout the problem, and the same solution procedure, as described for the uniform line, is employed. Naturally, account must be taken when formulating the equations at each node of the contribution from the capacitive stub. The use of stubs in this way offers added flexibility in dealing with nonuniformities or nonlinear components in complex networks. It is inevitable that the introduction of stubs will affect the propagation velocity of different frequency components and will thus introduce dispersion.

4.2 TLM MODELS FOR ONE-DIMENSIONAL ELECTROMAGNETIC PROBLEMS

Electromagnetic phenomena are described by Maxwell's equations.

$$\nabla \cdot \mathbf{D} = \rho \tag{4.23}$$

$$\nabla \cdot \mathbf{B} = 0 \tag{4.24}$$

$$\nabla x\mathbf{E} = -\frac{\partial \mathbf{B}}{\partial t} \tag{4.25}$$

$$\nabla x\mathbf{H} = \mathbf{j} + \frac{\partial \mathbf{D}}{\partial t} \tag{4.26}$$

In Cartesian coordinates the last two equations may be expanded as follows:

$$\begin{aligned} \frac{\partial E_z}{\partial y} - \frac{\partial E_y}{\partial z} &= -\frac{\partial B_x}{\partial t} \\ \frac{\partial E_x}{\partial z} - \frac{\partial E_z}{\partial x} &= -\frac{\partial B_y}{\partial t} \\ \frac{\partial E_y}{\partial x} - \frac{\partial E_x}{\partial y} &= -\frac{\partial B_z}{\partial t} \end{aligned} \tag{4.27}$$

$$\begin{aligned} \frac{\partial H_z}{\partial y} - \frac{\partial H_y}{\partial z} &= j_x + \frac{\partial D_x}{\partial t} \\ \frac{\partial H_x}{\partial z} - \frac{\partial H_z}{\partial x} &= j_y + \frac{\partial D_y}{\partial t} \\ \frac{\partial H_y}{\partial x} - \frac{\partial H_x}{\partial y} &= j_z + \frac{\partial D_z}{\partial t} \end{aligned} \tag{4.28}$$

For a one-dimensional problem, variations along the x-direction only are assumed. For propagation along $+x$, the only non-zero field components must be E_y and B_z. Based on these premises, Equations (4.27) and (4.28) simplify to

$$\frac{\partial E_y}{\partial x} = -\frac{\partial B_z}{\partial t} \tag{4.29}$$

$$-\frac{\partial H_z}{\partial x} = j_y + \frac{\partial D_y}{\partial t} \tag{4.30}$$

Multiplying Equation (4.30) by μ and applying Ohm's Law, $j_y = \sigma E_y$, where σ is the electrical conductivity of the medium, gives

$$-\frac{\partial B_z}{\partial x} = \mu\sigma E_y + \mu\varepsilon\frac{\partial E_y}{\partial t}$$

Taking the derivative of this equation with respect to t, differentiating Equation (4.29) with respect to x, and eliminating B_z gives

$$\frac{\partial^2 E_y}{\partial x^2} = \mu\varepsilon \frac{\partial^2 E_y}{\partial t^2} + \mu\sigma \frac{\partial E_y}{\partial t}$$

or, alternatively, in terms of the current density

$$\frac{\partial^2 j_y}{\partial x^2} = \mu\varepsilon \frac{\partial^2 j_y}{\partial t^2} + \mu\sigma \frac{\partial j_y}{\partial t} \tag{4.31}$$

An identical equation is obtained for B_z. These equations describe one-dimensional wave propagation in a lossy medium.

Equation (4.3), for the current in the general network discussed in the previous section, and for the case when $R = 0$, reduces to

$$\frac{\partial^2 i}{\partial x^2} = \frac{LC}{(\Delta x)^2} \frac{\partial^2 i}{\partial t^2} + \frac{GL}{(\Delta x)^2} \frac{\partial i}{\partial t} \tag{4.32}$$

Comparing (4.31) with (4.32) indicates the following equivalences:

$$i \leftrightarrow j_y \qquad \frac{C}{\Delta x} \leftrightarrow \varepsilon$$

$$\frac{L}{\Delta x} \leftrightarrow \mu \qquad \frac{G}{\Delta x} \leftrightarrow \sigma$$

Similarly, E_y may be associated with the voltage on the line [see Equation (4.4)]. Any general field excitation may be studied by defining the source voltage ${}_kV_s$ in the network problem. Parameter variations along the x-direction, such as changes in permittivity, permeability, or electrical conductivity, may be modeled by changing the capacitance C, inductance L, and conductance G of each line section. The only consideration that must be borne in mind is that the same propagation time must be maintained throughout the transmission line network so that synchronism is ensured. This may be elegantly achieved by using stubs as described earlier.

Solution of one-dimensional electromagnetic propagation problems can be obtained by first solving the network described in detail in the last section and then invoking the equivalence between circuit and field parameters.

4.3 STUDY OF DISPERSIVE EFFECTS IN ONE-DIMENSIONAL TLM MODELS

Discretization and the introduction of stubs introduce dispersion and, therefore, modeling errors. Discretization errors were examined in Section 2.3. In the present section, dispersion in a line made up of segments of the type shown in Fig. 4.7 is examined. Such a segment is the representation by TLM of a T-section with a total series inductance L and shunt capacitance C. Thus, the two link lines AB and BA′ in Fig. 4.7 represent the inductance and the stub BB′ the capacitance. Although this choice of model may not be the best available, it nevertheless illustrates velocity errors and dispersion.

For waves traveling in a periodic lossless structure of phase constant β,voltages and currents must be related by

$$\begin{bmatrix} \bar{V}_1 \\ \bar{I}_1 \end{bmatrix} = \begin{bmatrix} \cos(\beta\ell) & jZ\sin(\beta\ell) \\ j\dfrac{\sin(\beta\ell)}{Z} & \cos(\beta\ell) \end{bmatrix} \begin{bmatrix} \bar{V}_2 \\ \bar{I}_2 \end{bmatrix} \tag{4.33}$$

where the overbar indicates a phasor quantity. Quantities at the two ends of the line segment are related by the relevant transmission parameters for each section in Fig. 4.7. The segment shown in Fig. 4.7 is a cascade of three sections, namely AB, BB′ and BA′. Hence, the overall transmission matrix is

$$\mathbf{T} = \mathbf{T}_1\mathbf{T}_2\mathbf{T}_3 \tag{4.34}$$

where $[\mathbf{T}_i]$ is the matrix for section i.

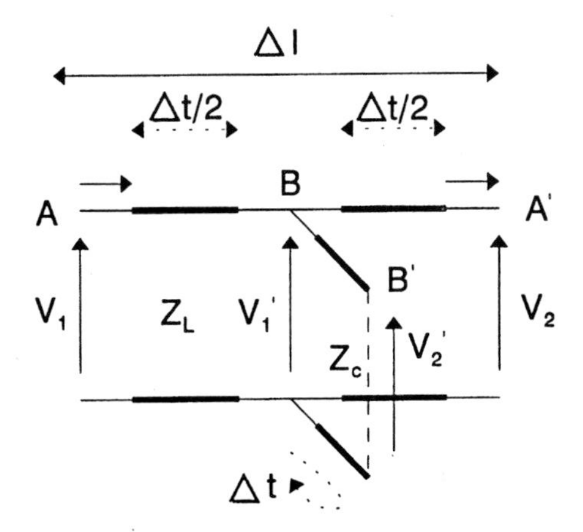

Fig. 4.7 Network used to study dispersion [$Z_L = L/\Delta t$, $Z_c = \Delta t/(2C)$]

Using the notation $\theta = \omega\Delta t/2$ and Equation (2.9), it is found that

$$\mathbf{T}_1 = \mathbf{T}_3 = \begin{bmatrix} \cos\theta & jZ_L\sin\theta \\ j\dfrac{\sin\theta}{Z_L} & \cos\theta \end{bmatrix} \tag{4.35}$$

The transmission matrix $[T_2]$ for the middle section, shown in more detail in Fig. 4.8, relates V_a, I_a to V_b, I_b; i.e.,

$$V_a = A'V_b + B'I_b$$

$$I_a = C'V_b + D'I_b \tag{4.36}$$

Since, however, $V_a = V_b$, it follows that $A' = 1$ and $B' = 0$, and from symmetry $D' = A' = 1$. The transmission equations for the two-port network defined by V'_1 and V'_2 are

$$V'_1 = \cos\theta V'_2$$

$$I'_1 = \frac{j\sin\theta V'_2}{Z_c} \tag{4.37}$$

where θ is equal to $\omega\Delta t/2$.

From Equation (4.36), the current I_a is equal to $C'V_b + I_b$, and from KCL it is equal to $I'_1 + I_b$. It therefore follows that

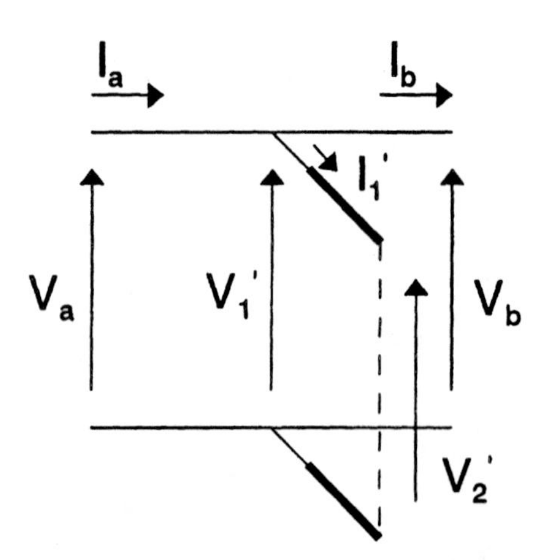

Fig. 4.8 Conditions in the middle section of Fig. 4.7

$$C' = \frac{I_1}{V_b} = \frac{I'_1}{V'_1}$$

and, using Equation (4.37), $C' = j\ tan\ \theta/Z_c$. The transmission matrix for the middle section is thus

$$\mathbf{T}_2 = \begin{bmatrix} 1 & 0 \\ j\dfrac{\tan\theta}{Z_c} & 1 \end{bmatrix} \tag{4.38}$$

Substituting Equations (4.35) and (4.38) into (4.34), the transmission parameters are obtained after some algebra, and after combining with (4.33) gives

$$\cos\left(\beta\ell\right) = 1 - 2\sin^2\theta\left[1 + \frac{Z_L}{2Z_c}\right] \tag{4.39}$$

The wave propagation velocity is obtained from Equation (4.39) after some algebra and is

$$u_w = u\frac{\pi\left(\dfrac{\Delta\ell}{\lambda_0}\right)}{\sin^{-1}\left[\left(1 + \dfrac{Z_L}{2Z_c}\right)^{1/2}\sin\left(\dfrac{\pi\Delta\ell}{\lambda_0}\right)\right]} \tag{4.40}$$

where u is the velocity of propagation on each line segment, and $\lambda_0 = u/f$ is the free-space wavelength. This expression shows, as expected, that the wave velocity depends on the ratio $\Delta\ell/\lambda_0$.

EXAMPLE 4.1 The significance of the dispersion relation of Equation (4.40) may be illustrated by means of a simple example showing the modeling of one-dimensional propagation in a medium with a propagation velocity equal to $1/\sqrt{\mu\varepsilon}$.

Let us assume that the velocity of propagation on each link line is chosen to be

$$u = \frac{\Delta\ell}{\Delta t} = 2\frac{1}{\sqrt{\mu\varepsilon}} \tag{4.41}$$

Clearly, the capacitive stubs will be required to slow down the wave to a speed appropriate for the medium considered. The link lines represent an inductance L equal to the inductance associated with the medium, while the associated link capacitance C_e is such that

$$\frac{1}{\sqrt{LC_e}} = 2\frac{1}{\sqrt{LC}} \tag{4.42}$$

where C is the capacitance associated with the medium. From Equation (4.42), the modeled capacitance is found to be $C_e = C/4$; hence the deficit, which must be modeled by the stub, is

$$C_s = C - C_e = 3\frac{C}{4}$$

Thus, $Z_L = L/\Delta t = \mu\Delta\ell/\Delta t$, and $Z_c = \Delta t/2C_s = \Delta t/(2 \times 0.75\ \varepsilon\ \Delta\ell)$, which results in a ratio $Z_L/2Z_c = 3$. Substituting in Equation (4.40) gives

$$\frac{u_w}{u} = \frac{\pi\left(\frac{\Delta\ell}{\lambda_0}\right)}{\sin^{-1}\left[2\sin\left(\frac{\pi\Delta\ell}{\lambda_0}\right)\right]} \tag{4.43}$$

Clearly, the expression in the square brackets must be smaller than one; therefore, $\Delta\ell \le \lambda_0/6$. Propagation cannot be sustained if the discretization length is chosen larger than $\lambda_0/6$. At this limiting value $u_w/u = 1/3$ and, since u was chosen to be equal to twice the medium wave velocity u_0 [see Equation (4.41)], it follows that $u_w/u_0 = 2/3$. This is a substantial velocity error which may not be acceptable. If a shorter discretization length is selected (e.g., $\Delta\ell/\lambda_0 = 0.1$), then, using Equation (4.43), it is found that $u_w/u_0 = 0.94$. This example illustrates the magnitude of velocity errors for particular choices of modeling parameters and the care required in constructing accurate numerical models. With the appropriate model, numerical dispersion is minimized, and TLM models may be used to study dispersion due to the physical parameters of the medium. Two configurations useful for studying plasma wave propagation and solitons are shown in Fig. 4.9a and b respectively [2–4].

■

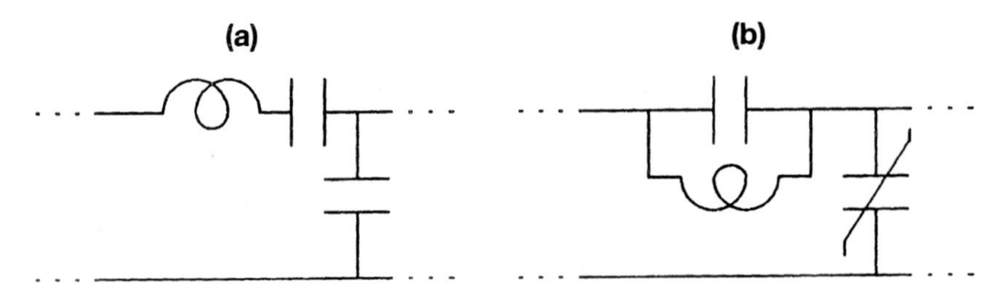

Fig. 4.9 Line sections used to study (a) plasma waves and (b) solitons

REFERENCES

[1] Boctor, S.A. 1987. *Electric Circuit Analysis.* Englewood Cliffs, NJ: Prentice-Hall.

[2] Hirose, A., and E.E. Lonngren. 1985. *Introduction to Wave Phenomena.* New York: John Wiley & Sons.

[3] Martin, D.J. 1980. Wave propagation in dispersive and attenuation media: A delay line simulation of plasma waves. *American Journal of Physics* 48, no. 6, 473-477.

[4] Giambo, S. et al. 1984. An electrical model for the Korteweg-de Vries equation. *American Journal of Physics* 52, no. 3, 238-243.

5

Two-Dimensional TLM Models

A whole class of problems in engineering and physics can be solved by using two-dimensional (2D) models. Two-dimensional problems are simpler to formulate and far less demanding on computational resources than three-dimensional (3D) models. They are therefore well suited for numerical experimentation and are used extensively even when it is anticipated that, in the final stage, a full 3D simulation will be necessary. Hence, it is important to study 2D modeling for its own sake but also to help understand the more complex 3D models.

In a 2D TLM model, waves propagate on a mesh of transmission lines interconnected at nodes as shown schematically in Fig. 5.1a. The precise configuration and parameters of these lines will be studied in detail in subsequent sections. It is, however, instructive to describe the essential features of the model without reference to any specific choice of transmission line configuration. A voltage pulse of 1 V magnitude is

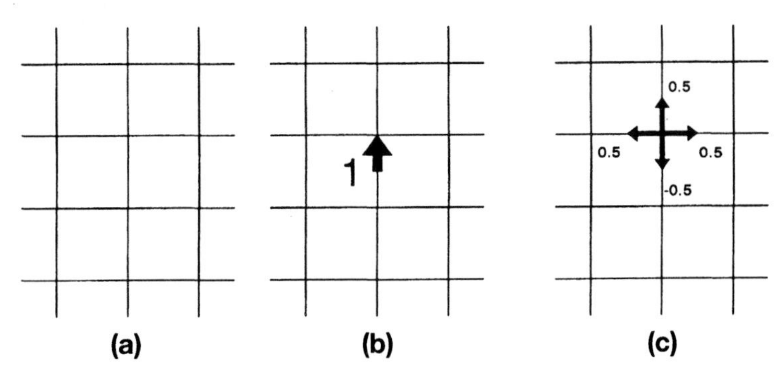

Fig. 5.1 Schematic of a TLM mesh: (a) impulse excitation and (b) results of the first scattering event.

shown incident on a node in Fig. 5.1b. This pulse will be partially reflected and transmitted according to transmission line theory. Assuming that each line has a characteristic impedance Z, then the incident pulse sees effectively three lines in parallel with an effective impedance $Z/3$. The reflection and transmission coefficient are $(Z/3 - Z)/(Z/3 + Z) = -0.5$, and $2(Z/3)/(Z + Z/3) = 0.5$, respectively. The scattered signals (reflected and transmitted) into the four lines are shown in Fig 5.1c. The energy injected into the lines by the incident pulse is $vi\Delta t = 1(1/Z)\Delta t = 1^2(\Delta t/Z)$. The energy of the scattered pulses is $[0.5^2 + 0.5^2 + 0.5^2 + (-0.5)^2](\Delta t/\mathrm{Z}) = 1^2(\Delta t/\mathrm{Z})$. Energy is thus conserved and spreads isotropically from the excited node. Similarly, the injected charge is $i\Delta t = (1/Z)\Delta t$t. Charge flow due to the scattered pulses is $(0.5 + 0.5 + 0.5 - 0.5)(\Delta t/Z) = (1/Z)\Delta t$. Charge is thus conserved. The -0.5 V scattered pulse combines with the incident pulse to produce a 0.5 V pulse, thus maintaining field continuity around the node. The scattered pulses propagate further to become incident on adjacent nodes and then scatter according to the rules described above. The disturbance thus propagates and spreads in the problem space. An example of several consecutive scattering events following an isotropic excitation is shown in Fig. 5.2. It can be seen from these figures that as each pulses impinges on a node it sets up a secondary spherical radiator. The waves emanating from several such radiators combine to form the overall waveform. This is in accordance with Huygens principle for light propagation [1]; the TLM model appears as a discrete implementation of this principle [2, 3].

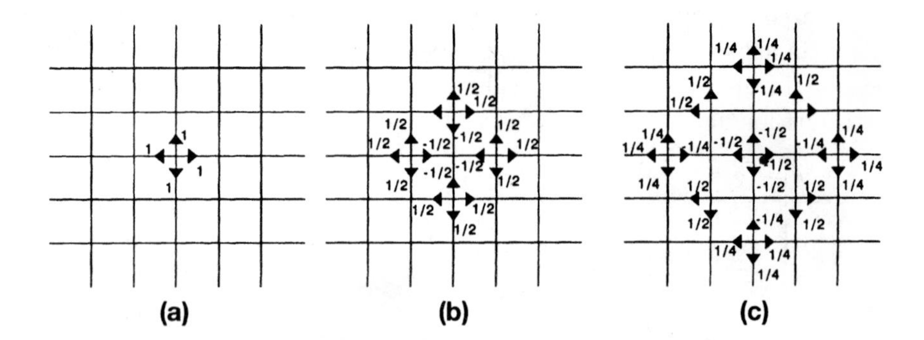

Fig. 5.2 Schematic of a TLM mesh: (a) impulse excitation and (b) results of the first scattering event.

In the next two sections, two nodal structures—described as the "series" and "shunt" nodes—used for 2D modeling are described in detail.

5.1 THE SERIES TLM NODE

Maxwell's equations in cartesian form were given in Section 4.2. Two field configurations on the x-y plane will be examined. The first admits an H_z component only (TE-modes), and the second an E_z component only (TM-modes). TE-modes are the subject of this section. TM-modes are discussed in section 5.2.

For TE-modes, the only non-zero field components are E_x, E_y, and H_z. No variations of field quantities along the z-direction are permitted. Maxwell's equations then reduce to:

$$\frac{\partial H_z}{\partial y} = \varepsilon \frac{\partial E_x}{\partial t} \tag{5.1}$$

$$-\frac{\partial H_z}{\partial x} = \varepsilon \frac{\partial E_y}{\partial t} \tag{5.2}$$

$$\frac{\partial E_y}{\partial x} - \frac{\partial E_x}{\partial y} = -\mu \frac{\partial H_x}{\partial t} \tag{5.3}$$

Differentiating Equations (5.1) and (5.2) with respect to y- and x- (respectively), adding the resulting equations, and combining with (5.3) eliminates the electric field component terms to produce the wave equations for 2D-propagation.

$$\frac{\partial^2 H_z}{\partial x^2} + \frac{\partial^2 H_z}{\partial y^2} = \mu\varepsilon \frac{\partial^2 H_z}{\partial t^2} \tag{5.4}$$

The next task is to devise two-dimensional circuits where the variation of circuit quantities (e.g., the current) are described by equations which are isomorphic to Equation (5.4).

A circuit structure, referred to as a "series node," describing a block of space of dimensions Δx, Δy, Δz is shown in Fig. 5.3 [4]. It consists of four ports as indicated. Intuitively, it is anticipated that the mesh current I, the x-directed voltages (ports 1, 3), and the y-directed voltages (ports 2, 4) may be related to the field quantities H_z, E_x, and E_y, respectively. To demonstrate this equivalence, conditions in the cluster of nodes shown in Fig. 5.4, part of a larger 2D cluster, will be examined in detail.

More specifically, conditions for the node indicated by the mesh current I_A are considered. The mesh currents I_x, I_y, I_z, and I_w are chosen equal to zero [5] and, hence, from Fig. 5.5a,

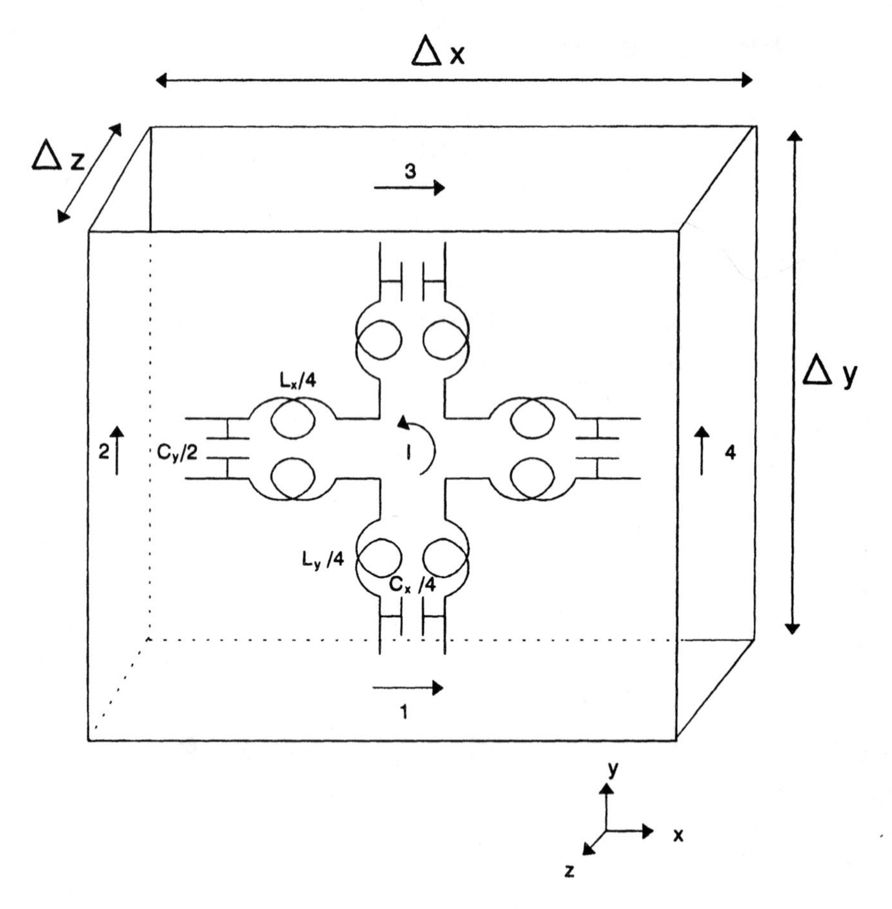

Fig. 5.3 The series TLM node

$$V_{AB} + V_{CD} + V_{EF} + V_{GH} = 2L\ \Delta\ell\ \frac{\partial I_A}{\partial t} \tag{5.5}$$

where Δx and Δy have been chosen equal to $\Delta\ell$. Hence,

$$-(V_{FE} - V_{AB}) + (V_{CD} - V_{HG}) = 2L\ \Delta\ell\ \frac{\partial I_A}{\partial t} \tag{5.6}$$

$$-\ [E_y(x + \Delta x) - E_y(x)]\,\Delta y + [E_x(y + \Delta y) - E_x(y)]\,\Delta x$$

$$= 2L\ \Delta\ell\ \frac{\partial I_A}{\partial t} \tag{5.7}$$

and dividing both sides by $\Delta x \Delta y$,

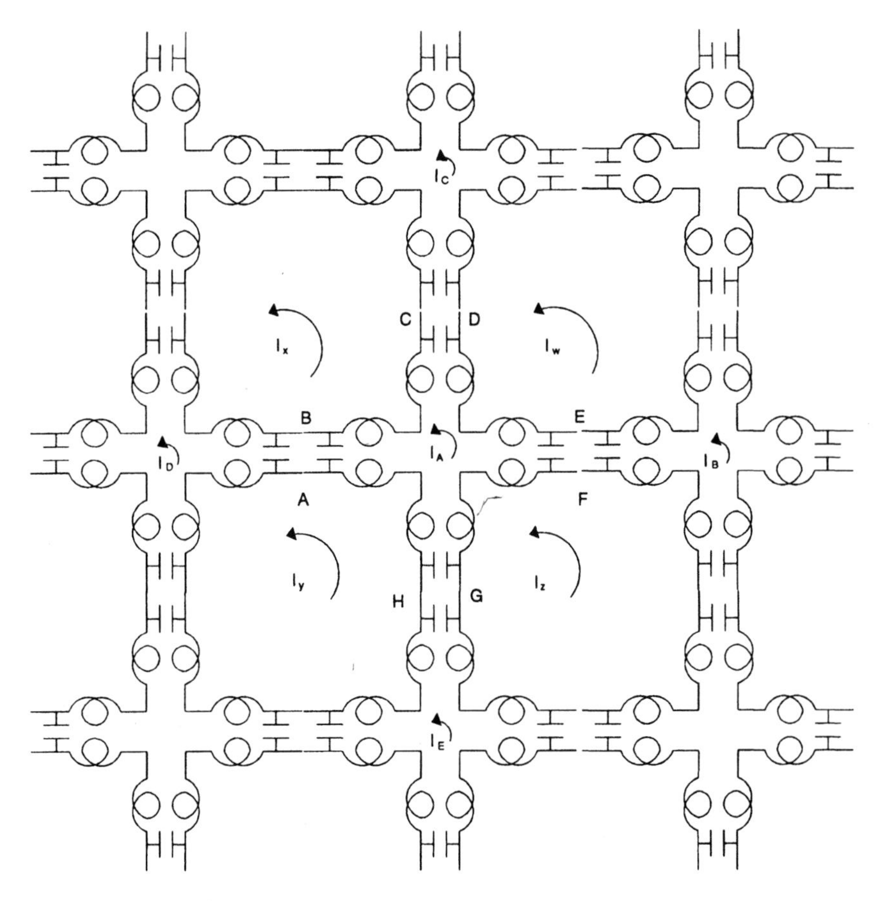

Fig. 5.4 A cluster of series nodes

$$-\frac{E_y(x+\Delta x)-E_y(x)}{\Delta x}+\frac{E_x(y+\Delta y)-E_x(y)}{\Delta y} = 2L\frac{\Delta \ell}{\Delta x \Delta y}\frac{\partial I_A}{\partial t}$$

Since $\Delta x = \Delta y = \Delta z = \Delta \ell$, and substituting $I_A = H_z\,\Delta z$ gives

$$\frac{\Delta E_x}{\Delta y}-\frac{\Delta E_y}{\Delta x} = 2L\frac{\partial H_z}{\partial t} \tag{5.8}$$

This equation reduces to one of Maxwell's equations as $\Delta \ell \to 0$ [see Equation (5.3)]. Similarly, from Fig. 5.5b, showing conditions across the capacitor connected between points E and F,

$$I_B - I_A = C\Delta x\frac{\partial V_{EF}}{\partial t}\text{, or} \tag{5.9}$$

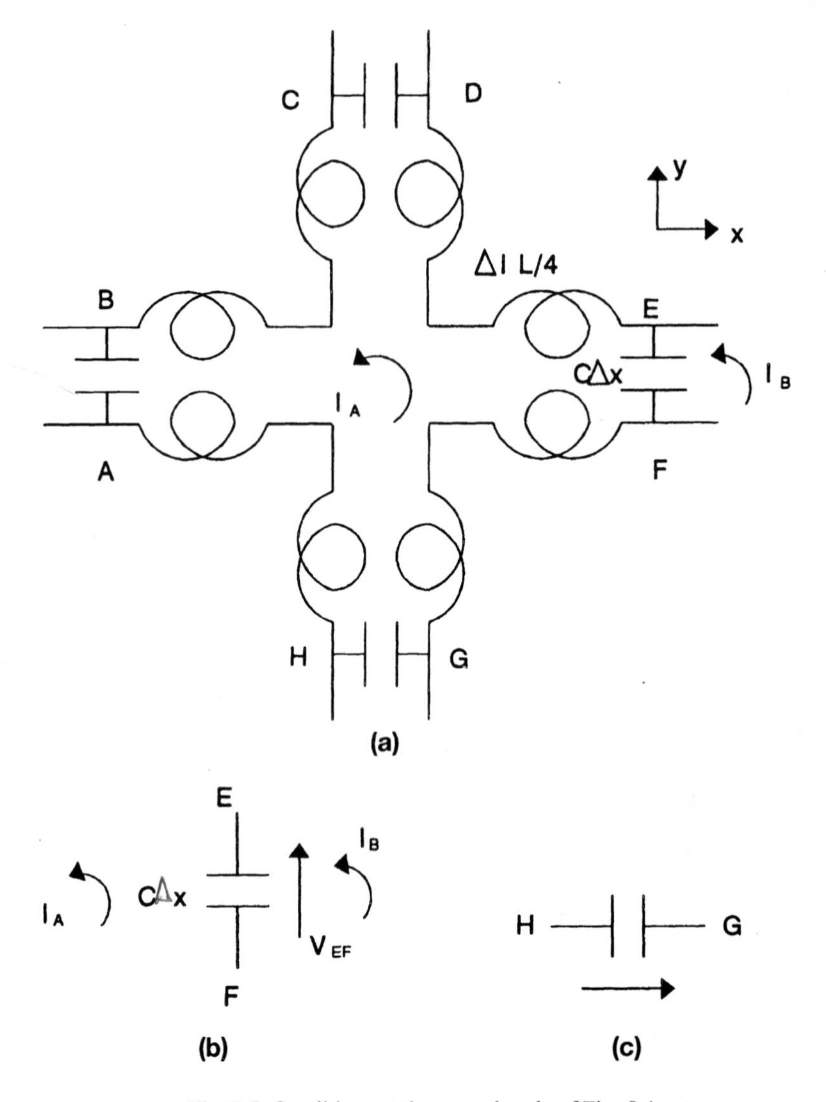

Fig. 5.5 Conditions at the central node of Fig. 5.4

$$\frac{[H_z(x+\Delta x) - H_z(x)]\,\Delta z}{\Delta x} = \frac{-C\Delta x \Delta y}{\Delta x}\frac{\partial E_y}{\partial t} \tag{5.10}$$

$$\frac{\partial H_z}{\partial x} = -C\frac{\partial E_y}{\partial t} \tag{5.11}$$

This equation should be compared with Equation (5.2). Conditions across the capacitor connected between points H and G, as shown in Fig. 5.5b, are described by an equation corresponding to Equation (5.1).

It remains now to relate the circuit parameters C_x, C_y, L_x, and L_y to the parameters of the medium being modeled. From Fig. 5.3, it is clear that

$$C_x = \varepsilon\frac{\Delta y \Delta z}{\Delta x},\ C_y = \varepsilon\frac{\Delta x \Delta z}{\Delta y}$$

The inductance L_x is associated with x-directed current I, producing a magnetic field intensity $H_z = I/\Delta z$ (Ampere's Law). The magnetic flux linked with this current is $\phi = \mu H_z (\Delta x\, \Delta y) = \mu I\, \Delta x\, \Delta y/\Delta z$. Hence,

$$L_x \equiv \frac{\phi}{I} = \mu\frac{\Delta x \Delta y}{\Delta z}$$

For y-propagation, the inductance L_y is similarly obtained and is

$$L_y = \mu\frac{\Delta x \Delta y}{\Delta z}$$

Since $\Delta x = \Delta y = \Delta z = \Delta\ell$, it follows that

$$L = \mu\Delta\ell$$

$$C = \varepsilon\Delta\ell \tag{5.12}$$

The equivalence between circuit and field quantities is then established from Equations (5.8) and (5.11) and is

$$\frac{I}{\Delta z} \leftrightarrow H_\ell \tag{5.13}$$

$$-\frac{V_y}{\Delta\ell} \leftrightarrow E_y \tag{5.14}$$

Similarly, from the circuit equivalent of Equation (5.1),

$$-\frac{V_x}{\Delta\ell} \leftrightarrow E_x \tag{5.15}$$

It therefore can be stated that the displayed circuit models TE-modes in a medium of parameters $(2\mu, \varepsilon)$. Intuitively, it can be seen that the loop current flows through the same inductance twice (in the x- and y- directions) and, therefore, inductance is overestimated by a factor of two. The manner in which circuit parameters are chosen to model any particular medium will be described in detail in Section 5.1.2. At this stage, it is sufficient to state that the series node may be represented by four transmission line segments each of a characteristic impedance Z_{TL}, as shown in Fig. 5.6. The subscript *TL* indicates that this is the characteristic impedance of each line and not that of the model or the medium. It is also assumed that $\Delta x = \Delta y = \Delta z = \Delta \ell$, and that the propagation time along each line is Δt. In the next section, the manner in which computation proceeds in a two-dimensional network of interconnected nodes, such as the one shown in Fig. 5.6, is described in detail.

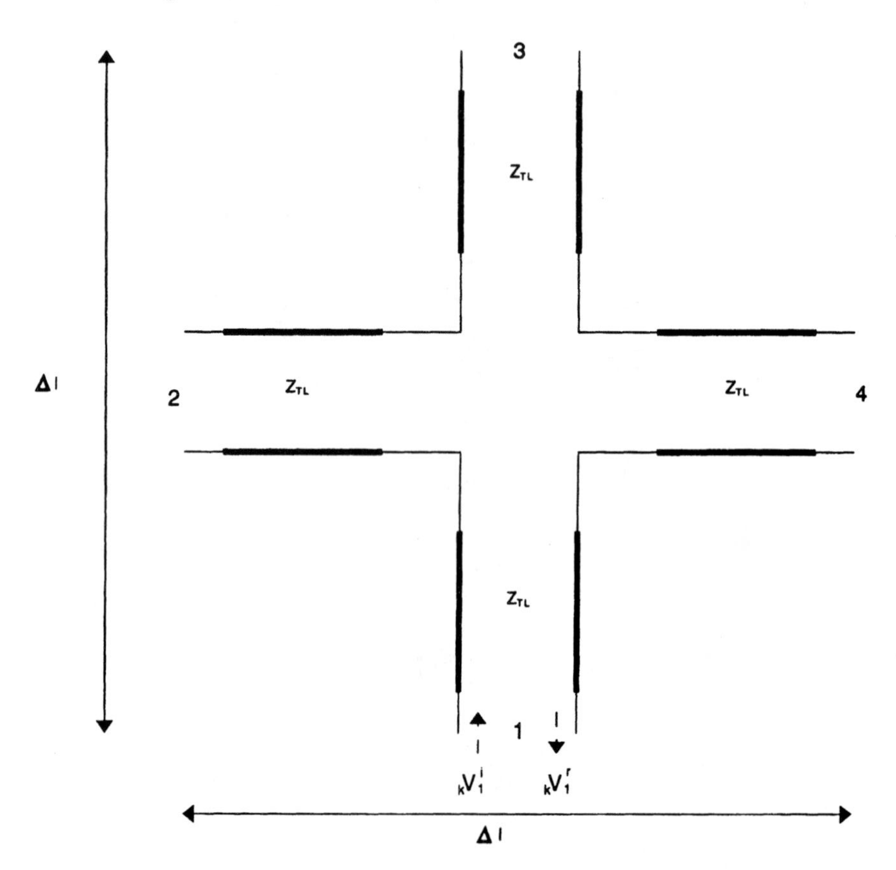

Fig. 5.6 Structure of the series node with four link line segments

5.1.1 Scattering and Computation in a Series Node

At time-step k, voltage pulses such as ${}_kV_1^i$ will be incident on the node as shown in Fig. 5.6. These pulses will be scattered according to transmission line theory, producing what are described as the reflected pulses shown in the same figure. The relationship between the incident ${}_kV^i$ and reflected ${}_kV^r$ pulses will be derived here using a general approach based on replacing each of the four line segments by its Thevenin equivalent. For each segment, this consists of a voltage source $2{}_kV^i$ in series with an impedance Z_{TL} as explained in Section 2.1. Conditions at the node can thus be determined from the equivalent circuit shown in Fig. 5.7, and the current is

$${}_kI = \frac{2{}_kV_1^i + 2{}_kV_4^i - 2{}_kV_3^i - 2{}_kV_2^i}{4Z_{TL}} \tag{5.16}$$

The field component H_z can then be obtained from

$$H_z = \frac{I}{\Delta\ell} = \frac{{}_kV_1^i - {}_kV_2^i - {}_kV_3^i + {}_kV_4^i}{4\Delta\ell Z_{TL}} \tag{5.17}$$

The electric field components can be obtained from Equations (5.14) and (5.15) after calculating the average value of the voltage in the x- and y-directions, namely

$${}_kE_x = -\frac{{}_kV_1^i + {}_kV_3^i}{\Delta\ell} \tag{5.18}$$

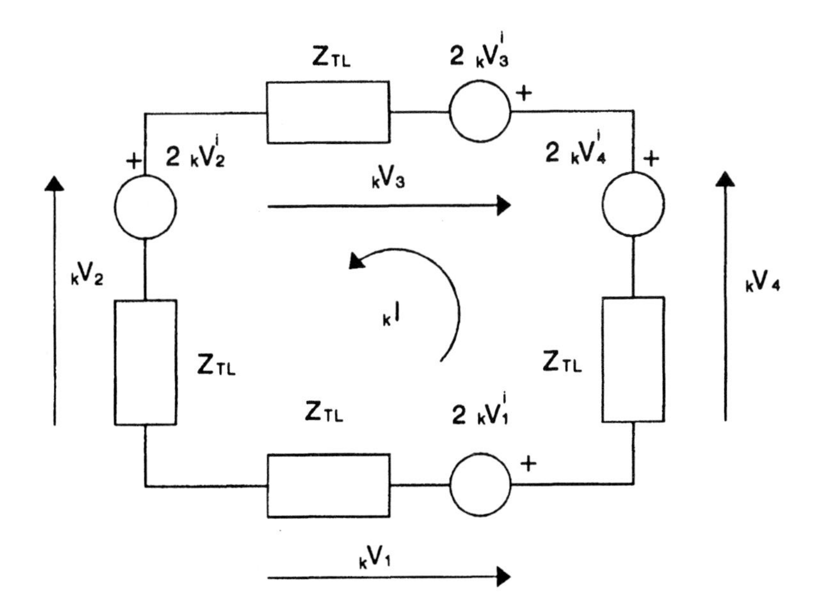

Fig. 5.7 Thevenin equivalent circuit of the series node

$$_kE_y = -\frac{_kV_2^i + {}_kV_4^i}{\Delta\ell} \tag{5.19}$$

Equations (5.17) through (5.19) indicate how the field components are derived from the incident voltages.

The reflected voltages can be obtained easily from the general expression

$$_kV^r = {}_kV - {}_kV^i \tag{5.20}$$

where $_kV$ is the total voltage at time-step k. Applying Equation (5.20) to find the reflected voltage on port 1 gives

$$_kV_1^r = {}_kV_1 - {}_kV_1^i = 2\,{}_kV_1^i - I\,Z_{TL} - {}_kV_1^i = {}_kV_1^i - I\,Z_{TL}$$

Substituting the current from Equation (5.16) gives

$$_kV_1^r = 0.5\left({}_kV_1^i + {}_kV_2^i + {}_kV_3^i - {}_kV_4^i\right)$$

Similar expressions are obtained for the other reflected voltages so that the vector of reflected voltages $_k\mathbf{V}^r$ is related to the vector of incident voltages $_k\mathbf{V}^i$ by a scattering matrix $\mathbf{S}$; i.e.,

$$_k\mathbf{V}^r = \mathbf{S}\,{}_k\mathbf{V}^i \tag{5.21}$$

This expression, written in full for the series node, is given below

$$\begin{bmatrix} _kV_1^r \\ _kV_2^r \\ _kV_3^r \\ _kV_4^r \end{bmatrix} = 0.5 \begin{bmatrix} 1 & 1 & 1 & -1 \\ 1 & 1 & -1 & 1 \\ 1 & -1 & 1 & 1 \\ -1 & 1 & 1 & 1 \end{bmatrix} \begin{bmatrix} _kV_1^i \\ _kV_2^i \\ _kV_3^i \\ _kV_4^i \end{bmatrix} \tag{5.22}$$

In practical implementations, it is more efficient to calculate the current and then obtain the reflected voltages as shown earlier. It is also possible to use a single storage location for both incident and reflected voltages at each port.

It remains to show how the incident voltages at the next time-step $k + 1$ may be obtained. This step in the computation may be understood by reference to the cluster of nodes shown schematically in Fig. 5.8.

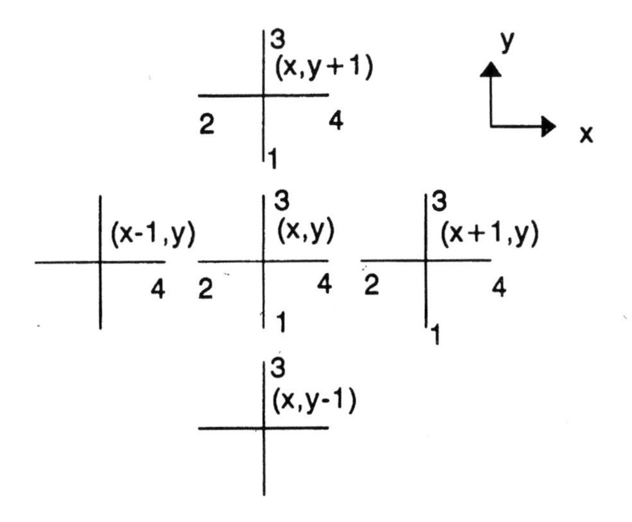

Fig. 5.8 Connection in a mesh of series nodes

Pulses incident on ports of node (x, y) at time step $k + 1$ clearly are pulses reflected from adjacent nodes at the previous time-step k. Hence, the new incident voltages on node (x, y) depend entirely on the nodes connected to it. From Fig. 5.8, the following expressions are obtained by inspection:

$$
\begin{aligned}
{}_{k+1}V_1^i(x, y) &= {}_kV_3^r(x, y-1) \\
{}_{k+1}V_2^i(x, y) &= {}_kV_4^r(x-1, y) \\
{}_{k+1}V_3^i(x, y) &= {}_kV_1^r(x, y+1) \\
{}_{k+1}V_4^i(x, y) &= {}_kV_2^r(x+1, y)
\end{aligned}
\tag{5.23}
$$

Equations (5.23) describe the "connection" process that permits the calculation of the incident voltages at the new time step. The current at time step $k + 1$ may be obtained from Equation (5.16), where the new incident voltages ${}_{k+1}V^i$ now are used. The process may be repeated for as many time steps as desired. At any time, the field quantities may be obtained at any node using Equations (5.17) through (5.19).

The calculation commences at $k = 0$ by imposing the initial conditions. To excite electric field component E_y at node (x_0, y_0) it is necessary to define incident voltages ${}_0V_2^i(x_0, y_0)$ and ${}_0V_4^i(x_0, y_0)$, as Equation (5.19) confirms. All other incident voltages at $k = 0$ are then set to zero. The computation procedure is shown schematically in Fig. 5.9.

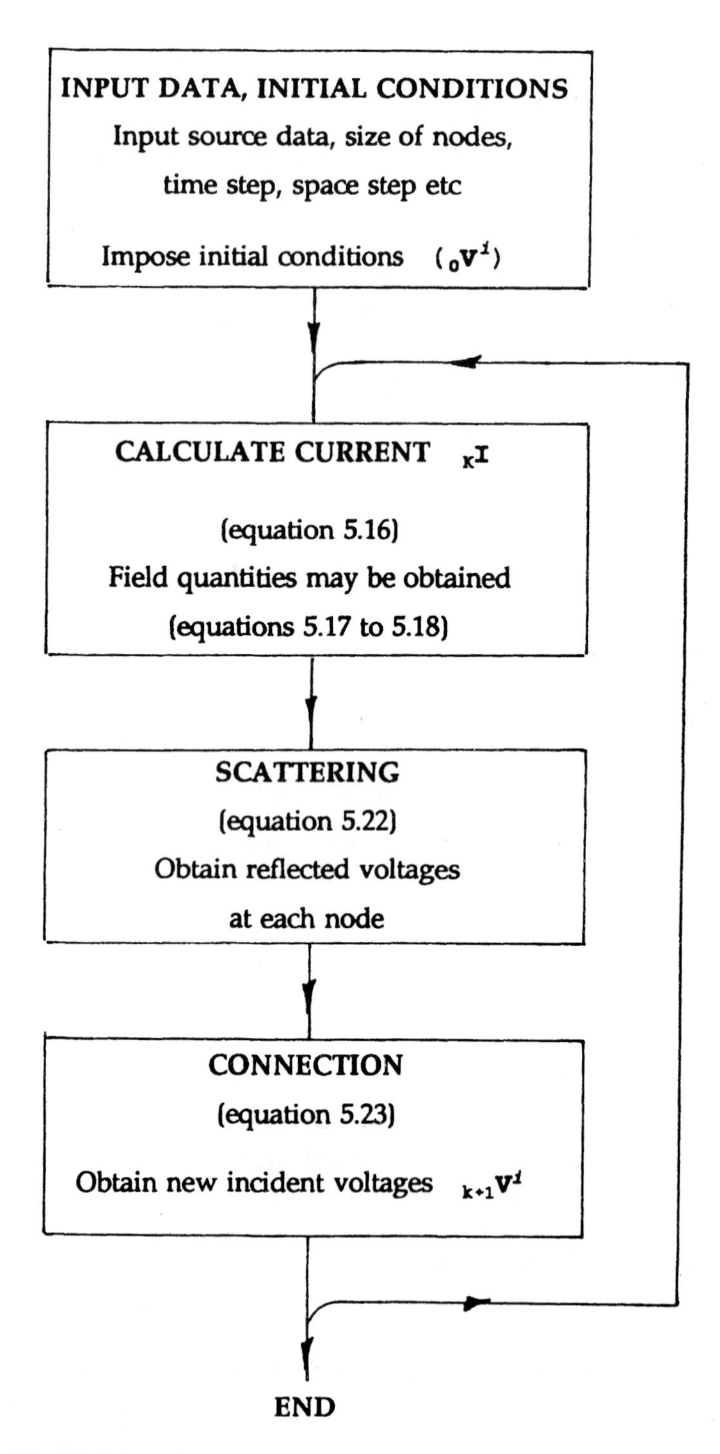

Fig. 5.9 Schematic of the computational procedure in a series node mesh

In practical problems, there will be boundaries—such as conducting boundaries—where the connection procedure described by Equation (5.23) has to be modified. On a port connected to a perfectly conducting boundary, the reflected pulse is returned with opposite polarity to become the incident pulse at the next time-step. A thorough discussion of boundary conditions is presented in the next section.

5.1.2 Modeling of Inhomogeneous Lossy Materials and Boundaries

Solving problems where different materials are present needs modifications to the basic approach described in the previous section. Similarly, losses must be incorporated into the model and different types of boundary, physical or numerical, must be adequately described. These issues are addressed separately.

Modeling of Inhomogeneous Media

On first reflection, the modeling of different materials requires only the adjustment of the circuit parameters (C and L) to account for the local value of dielectric permittivity ε and magnetic permeability μ. The immediate consequence is that the velocity of propagation is not the same throughout the problem and is given by the formula

$$u = \frac{1}{\sqrt{\mu\varepsilon}}$$

Two choices appear to be possible. First, the same time-step may be maintained throughout the problem, thus resulting in a space discretization length that varies depending on u. Second, the same space discretization length may be maintained throughout the problem, thus resulting in a time-step that varies depending on u. In the former case, connectivity between ports on either side of the interface between different materials is not maintained (because $\Delta\ell$ is different), in contrast to the situation in 1D models. In the latter case, synchronism is lost because pulses arrive at a different rate from either side of the interface (because Δt is not the same). Therefore, a different approach is necessary whereby both connectivity and synchronism are maintained. This can be achieved by introducing, whenever necessary, extra inductance (to obtain the desired value of μ) and extra capacitance (to obtain ε) in the form of stubs.

The magnetic permeability may be increased by adding an inductance L_s in a series node as shown in Fig. 5.10a. Synchronism is maintained by choosing the round-trip time for the stub representing L_s to be the same as for the transmission lines forming the rest of the node. The series node with an inductive stub is shown in Fig 5.10b, where port 5 is the connection to the stub. Each line has a characteristic impedance Z_{TL}, and the stub has an impedance Z_s. The Thevenin equivalent circuit may be constructed as before, and it is shown in Fig. 5.10c. The loop current is

$$_kI = \frac{2_kV_1^i + 2_kV_4^i - 2_kV_3^i - 2_kV_2^i + 2_kV_s^i}{4Z_{TL} + Z_s} \tag{5.24}$$

and all other parameters are calculated as described in Section 5.1.1. The reflected voltage at each port is evaluated as before, using Equation (5.20), to give the following matrix equation relating reflected and incident voltages

$$\begin{bmatrix} _kV_1^r \\ _kV_2^r \\ _kV_3^r \\ _kV_4^r \\ _kV_5^r \end{bmatrix} = \frac{1}{Z}\begin{bmatrix} Z-2 & 2 & 2 & -2 & -2 \\ 2 & Z-2 & -2 & 2 & 2 \\ 2 & -2 & Z-2 & 2 & 2 \\ -2 & 2 & 2 & Z-2 & -2 \\ -2Z_s/Z_{TL} & +2Z_s/Z_{TL} & +2Z_s/Z_{TL} & -2Z_s/Z_{TL} & 4-Z_s/Z_{TL} \end{bmatrix}\begin{bmatrix} _kV_1^i \\ _kV_2^i \\ _kV_3^i \\ _kV_4^i \\ _kV_5^i \end{bmatrix} \tag{5.25}$$

where $Z = 4 + Z_s/Z_{TL}$.

Computation then proceeds as for the case of a mesh of nodes without stubs, but with the scattering matrix given by Equation (5.25).

Series Node with Losses

Losses may be incorporated into the model by introducing a resistance R_s as shown in Fig. 5.11. This resistor may be viewed as a matched transmission line of characteristic impedance R_s. The Thevenin equivalent circuit shown in Fig. 5.10c applies, but with an additional series component R_s. Since the lossy stub is matched, no pulse from it becomes incident onto the node. Energy is simply extracted (converted to heat) by this stub from the node. The scattering matrix for the case when both an inductive stub and losses are present is

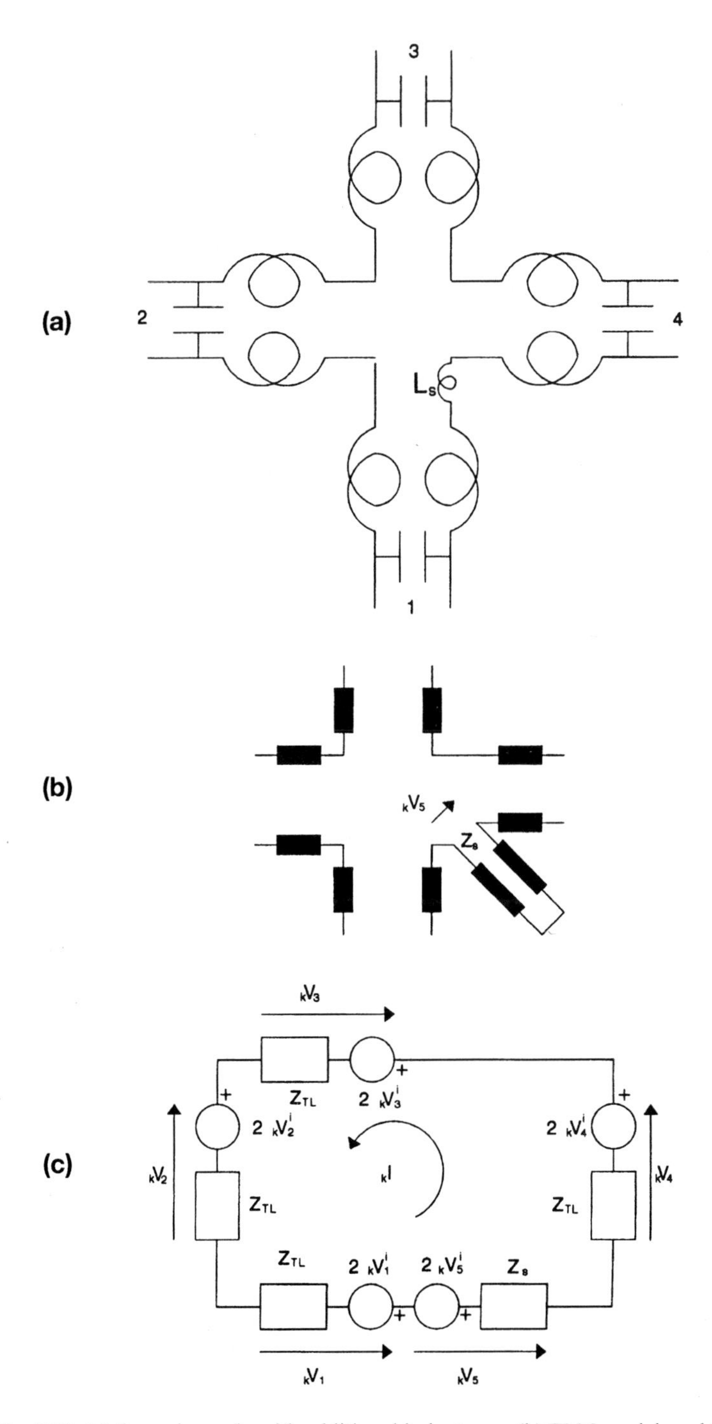

Fig. 5.10 (a) the series node with additional inductance, (b) TLM model, and (c) Thevenin equivalent

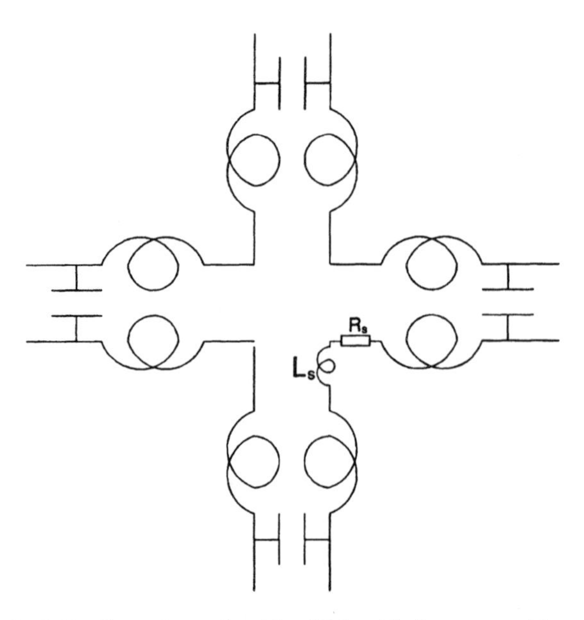

Fig. 5.11 The series node with additional inductance and losses

$$\mathbf{S} = \frac{1}{Z}\begin{bmatrix} Z-2 & 2 & 2 & -2 & -2 \\ 2 & Z-2 & -2 & 2 & 2 \\ 2 & -2 & Z-2 & 2 & 2 \\ -2 & 2 & 2 & Z-2 & -2 \\ 2\hat{Z}_s & 2\hat{Z}_s & 2\hat{Z}_s & -2\hat{Z}_s & Z-2\hat{Z}_s \end{bmatrix} \tag{5.26}$$

where

$$\hat{Z}_s = Z_s/Z_{TL}$$

$$\hat{R}_s = R_s/Z_{TL}$$

$$Z = 4 + \hat{Z}_s + \hat{R}_s$$

The caret symbol (^) indicates a quantity normalized to Z_{TL}. A more general approach, using a hybrid method, for modeling permittivity, permeability, and losses in a series node is described in Ref. [6].

Choice of Model Parameters

The choice of model parameters appropriate for describing a medium of parameters (μ, ε) is now considered. Let each transmission line in a node have total capacitance and inductance of C and L, respectively. The characteristic impedance of each transmission line is

$$Z_{TL} = \sqrt{L_d/C_d} = \sqrt{L/C} \tag{5.27}$$

Similarly, the velocity of propagation on each transmission line is

$$u_{TL} = \frac{1}{\sqrt{L_d C_d}} = \frac{\Delta\ell}{\sqrt{LC}}$$

and hence,

$$\Delta t = \frac{\Delta\ell}{u_{TL}} = \sqrt{LC} \tag{5.28}$$

From Equations (5.27) and (5.28), it follows that

$$L = Z_{TL}\Delta t$$

$$C = \frac{\Delta t}{Z_{TL}} \tag{5.29}$$

The subscript TL indicates parameters of each transmission line and not of the TLM model or the medium.

EXAMPLE 5.1 *Choice of modeling parameters for a medium of parameters* (ε_0, μ_0). The velocity of propagation and the characteristic impedance in each transmission line are chosen to be

$$u_{TL} = \frac{\Delta\ell}{\Delta t} = \sqrt{2}\,\frac{1}{\sqrt{\mu_0\varepsilon_0}} = \sqrt{2}c_0$$

$$Z_{TL} = \frac{1}{\sqrt{2}}\sqrt{\frac{\mu_0}{\varepsilon_0}} = \frac{Z_0}{\sqrt{2}}$$

It remains to show that this choice models the correct amount of inductance and capacitance.

From Equation (5.29),

$$L = Z_{TL}\Delta t = \frac{1}{\sqrt{2}}\sqrt{\frac{\mu_0}{\varepsilon_0}}\,\frac{\sqrt{\mu_0\varepsilon_0}\Delta\ell}{\sqrt{2}} = \mu_0\frac{\Delta\ell}{2}$$

Since the mesh current flows in an inductance $2L$, the effective inductance modeled is $2(\mu_0\,\Delta\ell/2) = \mu_0\,\Delta\ell$, as desired. Similarly, the modeled capacitance is

$$C = \frac{\Delta t}{Z_{TL}} = \frac{\sqrt{\mu_0 \varepsilon_0}\Delta \ell}{\sqrt{2}} \frac{\sqrt{2\varepsilon_0}}{\sqrt{\mu_0}} = \varepsilon_0 \Delta \ell$$

Since each voltage pulse encounters capacitance C, this is effectively the modeled capacitance and has the desired value. The node and its parameters are shown in Fig. 5.12.

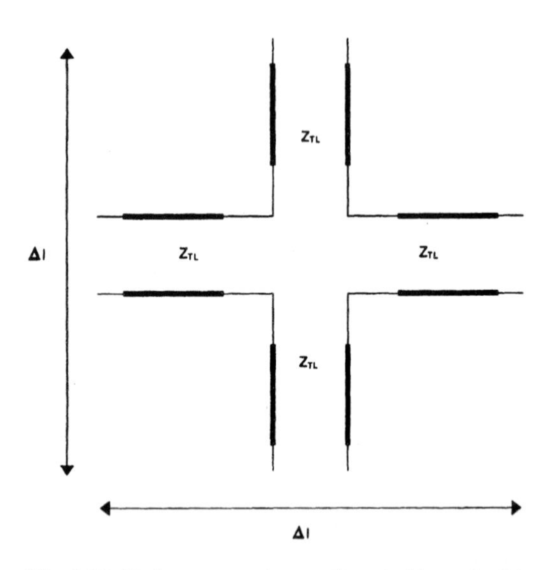

Fig. 5.12 Node representing medium in Example 5.1

■

EXAMPLE 5.2 *Choice of modeling parameters for a medium of parameters* (ε_0, μ). The velocity of propagation and the characteristic impedance of each transmission line are chosen to be:

$$u_{TL} = \frac{\Delta \ell}{\Delta t} = \sqrt{2} \frac{1}{\sqrt{\mu_0 \varepsilon_0}}$$

$$Z_{TL} = \frac{1}{\sqrt{2}} \sqrt{\frac{\mu_0}{\varepsilon_0}}$$

From Example 5.1, it is evident that this choice describes capacitance and inductance equal to $\varepsilon_0 \Delta \ell$ and $\mu_0 \Delta \ell$, respectively. While the capacitance has the desired value, the inductance is smaller than it should be by an amount $L_s = \mu \Delta \ell - \mu_0 \Delta \ell = (\mu_r - 1) \mu_0 \Delta \ell$. This inductance is added in the form of a stub of characteristic impedance $Z_s = 2L_s/\Delta t$. The node and its parameters are shown in Fig. 5.13.

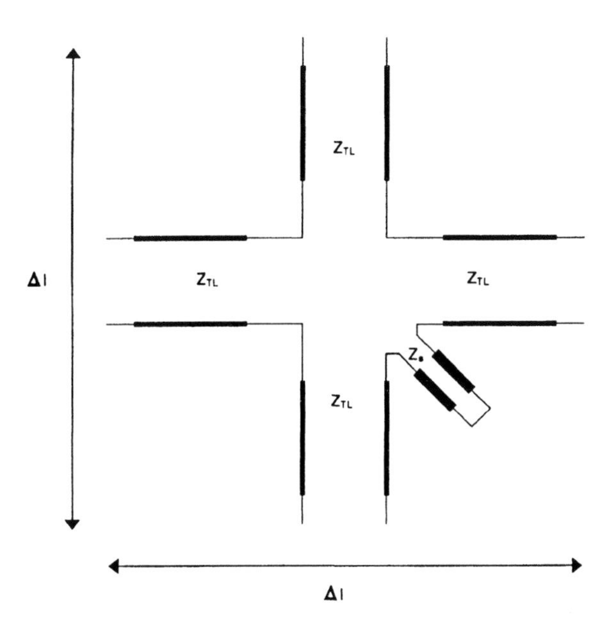

Fig. 5.13 Node representing medium in Example 5.2

■

Modeling of Boundaries

In every numerical simulation, it is necessary to describe boundaries. The most common case is perfectly conducting boundaries, which may be the external boundaries of the problem or internal boundaries describing conductors placed inside the volume described by the simulation. Conducting boundaries may be inserted at the node or between nodes as shown in Fig. 5.14. The "short circuit node" shown in Fig. 5.14a can be described by modifying the scattering matrix [see Equation (5.22)] to take account that at each port $V^r = -V^i$. The short-circuit boundary shown in Fig. 5.14b is simply described by recognizing that during the connection process

$$ {}_{k+1}V_4^i = -{}_kV_4^r $$

An open-circuit boundary describes effectively a plane of symmetry. This can be seen from Fig. 5.15a where it is assumed that there is symmetry with respect to the plane passing from 00′. Symmetry implies that at all times

$$ {}_kV_4^r(x,y) = {}_kV_2^r(x+1,y) $$

Hence, the new incident voltage on node (x, y) is

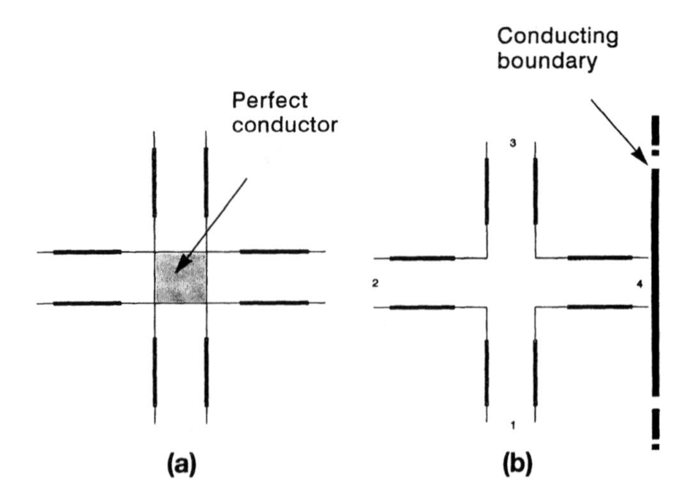

Fig. 5.14 Representation of (a) conductors and (b) conducting boundaries

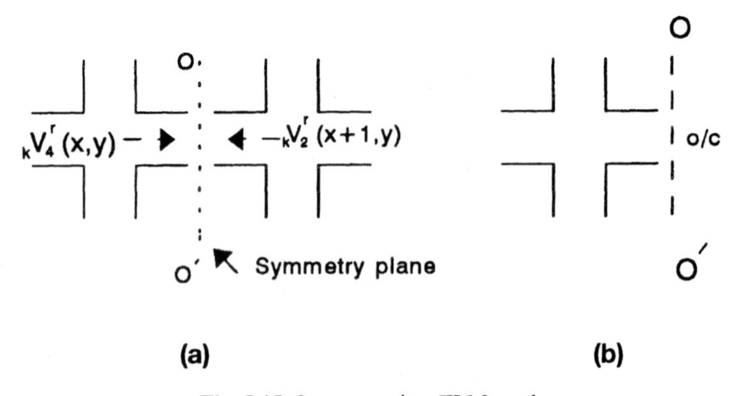

Fig. 5.15 Symmetry in a TLM mesh

$$_{k+1}V_4^i\,(x, y) \;=\; {}_kV_2^r\,(x+1, y) \;=\; {}_kV_4^r\,(x, y)$$

Hence, the problem to the right of line 00′ need not be modeled in detail, and it suffices to insert at 00′ an open-circuit boundary condition as shown in Fig. 5.15b.

A more difficult problem is the description of open boundaries (boundary at infinity) in a numerical simulation. As already mentioned, a numerical boundary is required to reduce the problem to a finite size for solution by computer. For this purpose a "matched" (absorbing) boundary condition is used in TLM as shown in Fig 5.16. The impedance Z connected to port 4 is chosen to be equal to the characteristic impedance of the medium. For the case of Example 5.1, $Z = Z_0$; hence, the reflection coefficient on port 4 is

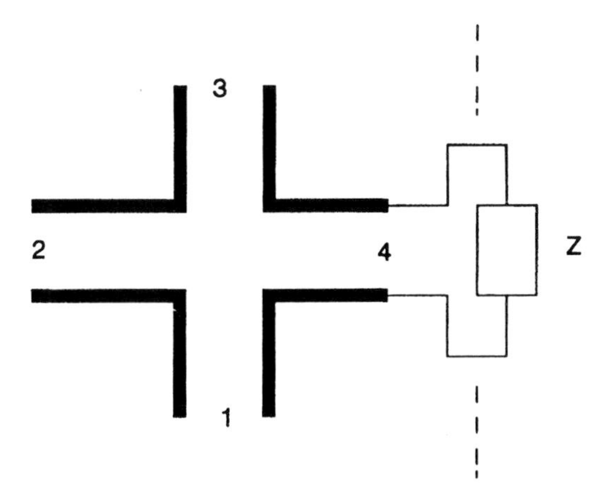

Fig. 5.16 Termination by a boundary of impedance Z

$$\Gamma = \frac{Z - Z_{TL}}{Z + Z_{TL}} = \frac{Z_0 - \dfrac{Z_0}{\sqrt{2}}}{Z_0 + \dfrac{Z_0}{\sqrt{2}}} = \frac{\sqrt{2} - 1}{\sqrt{2} + 1}$$

In most cases, this approach gives good results. A worthwhile improvement is to adjust the reflection coefficient to take account of the angle of incidence of the wave on the boundary [7]. For a wave incident on a boundary at $x = x_0$ at an angle ϕ, as shown in Fig. 5.17, $E_z = E_0$, $H_x = H_0 \sin \phi$, and $H_y = -H_0 \cos \phi$. The medium impedance is $E_0/H_0 = Z_0$. The wave impedance in the x-direction is then

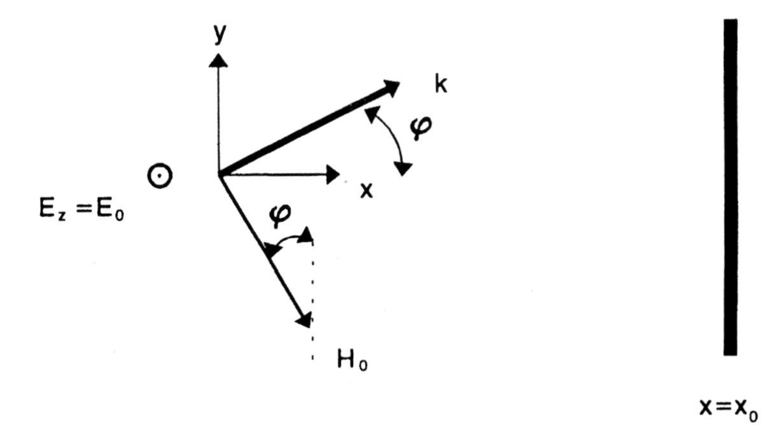

Fig. 5.17 A wave incident on a boundary at an angle ϕ

$$Z_x = \frac{E_z}{-H_y} = \frac{E_0}{-(-H_0 \cos\phi)} = \frac{Z_0}{\cos\phi}$$

For an absorbing boundary condition, the x-directed lines must then be terminated by Z_x. Further improvements can be made, at the expense of increased computational effort, by using a Green's function approach for the description of boundaries as described in Refs. [8, 3].

5.2 THE SHUNT TLM NODE

Following a procedure similar to that of Section 5.1, Maxwell's equations reduce to the following set for the case of TM-modes (field components H_x, H_y, and E_z).

$$\frac{\partial E_z}{\partial y} = -\mu \frac{\partial H_x}{\partial t} \tag{5.30}$$

$$-\frac{\partial E_z}{\partial x} = -\mu \frac{\partial H_y}{\partial t} \tag{5.31}$$

$$\frac{\partial H_y}{\partial x} - \frac{\partial H_x}{\partial y} = \varepsilon \frac{\partial E_z}{\partial t} \tag{5.32}$$

Differentiating Equations (5.30) and (5.31) with respect to y- and x-, respectively, adding the resulting equations and combining with (5.32) to eliminate the magnetic field components gives

$$\frac{\partial^2 E_z}{\partial x^2} + \frac{\partial^2 E_z}{\partial y^2} = \mu\varepsilon \frac{\partial^2 E_z}{\partial t^2} \tag{5.33}$$

This is the wave equation for 2D propagation. The next task is to devise a suitable circuit wherein the variation of circuit quantities (e.g., voltage) are described by equations that are isomorphic to Equations (5.30) through (5.33).

A circuit structure, referred to as a "shunt node," describing a block of space of dimensions Δx, Δy, and Δz is shown in Fig. 5.18 [4, 9]. Intuitively, it is expected that the voltage V_z and the current into each port may be related to E_z, H_x, and H_y. The relationship between field and circuit parameters may be established by methods similar to those used in Section 5.1 [5]. However, an alternative, more intuitive approach is adopted here. Propagation along the x- and y-directions is studied separately, and

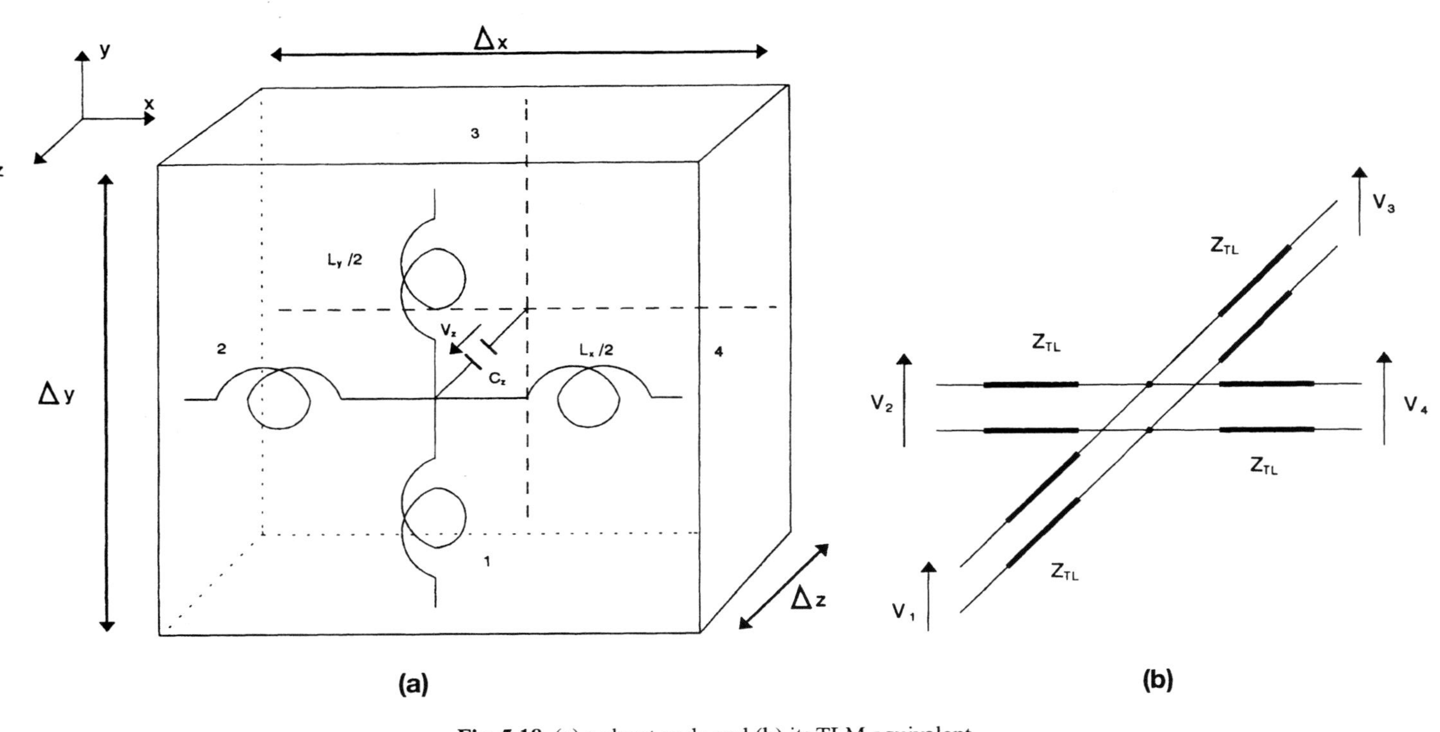

Fig. 5.18 (a) a shunt node and (b) its TLM equivalent

the results are combined to produce an equation describing the spatial and temporal variation of V_z.

For x-propagation, the effective circuit is shown in Fig. 5.19a. From Kirchhoff's voltage and current laws,

$$L_x \frac{\partial I_x}{\partial t} = -\frac{\partial V_z}{\partial x}\Delta x \tag{5.34}$$

$$C_z \frac{\partial V_z}{\partial t} = -\frac{\partial I_x}{\partial x}\Delta x \tag{5.35}$$

Differentiating the last two equations with respect to x and t, respectively, and combining gives

$$\frac{\partial^2 V_z}{\partial x^2}\frac{(\Delta x)^2}{L_x} = C_z \frac{\partial^2 V_z}{\partial t^2} \tag{5.36}$$

Similarly, for y-propagation (Fig. 5.19b),

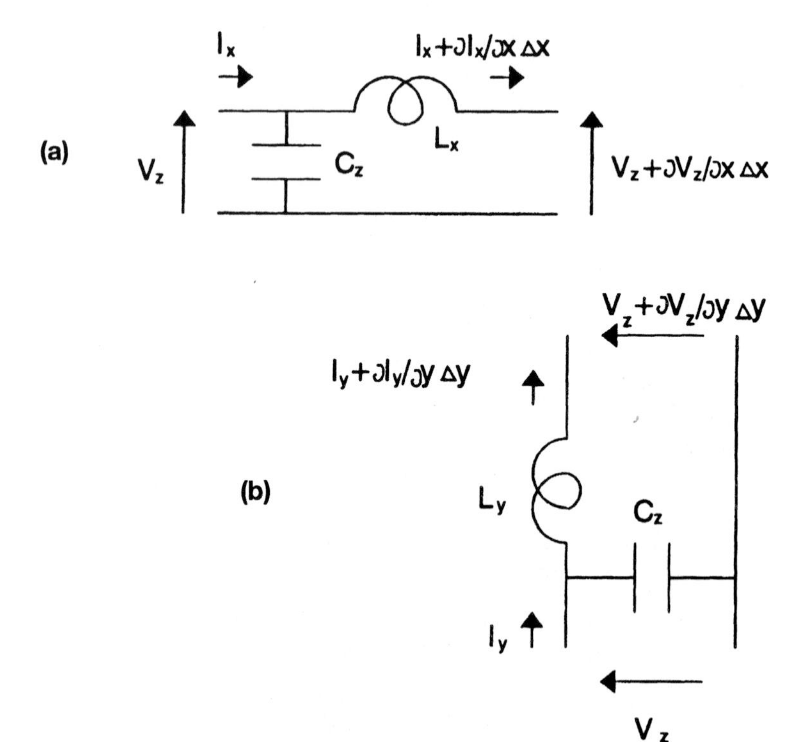

Fig. 5.19 Conditions at the shunt node for propagation in (a) the x and (b) the y directions

$$\frac{\partial^2 V_z}{\partial y^2}\frac{(\Delta y)^2}{L_y} = C_z\frac{\partial^2 V_z}{\partial t^2} \tag{5.37}$$

Combining (5.36) and (5.37)

$$\frac{\partial^2 V_z}{\partial x^2}\frac{(\Delta x)^2}{L_x} + \frac{\partial^2 V_z}{\partial y^2}\frac{(\Delta y)^2}{L_y} = 2\,C_z\frac{\partial^2 V_z}{\partial t^2} \tag{5.38}$$

The circuit parameters may be calculated as for the series node.

$$L_x = \mu\frac{\Delta x\Delta z}{\Delta y},\ L_y = \mu\frac{\Delta y\Delta z}{\Delta x},\ C_z = \varepsilon\frac{\Delta x\Delta y}{\Delta z} \tag{5.39}$$

Substituting (5.39) into (5.38) gives

$$\frac{\partial^2}{\partial x^2}\left(\frac{V_z}{\Delta z}\right) + \frac{\partial^2}{\partial y^2}\left(\frac{V_z}{\Delta z}\right) = (2\varepsilon)\,\mu\frac{\partial^2}{\partial t^2}\left(\frac{V_z}{\Delta z}\right) \tag{5.40}$$

This equation may be compared with the wave equation for TM-modes [see Equation (5.33)] to establish the following equivalences:

$$-\frac{V_z}{\Delta z} \leftrightarrow E_z$$

$$\frac{I_x}{\Delta y} \leftrightarrow H_y$$

$$-\frac{I_y}{\Delta x} \leftrightarrow H_x \tag{5.41}$$

where the last two relationships are derived by comparing Equations (5.30) through (5.32) with (5.34) and (5.35).

The shunt node, with the parameters chosen, models a medium of parameters $(2\varepsilon, \mu)$.

5.2.1 Scattering and Computation in a Shunt Node

The derivation of the scattering properties of this node is done in exactly the same way as for the series node. Observing from the center of the node toward each port, and replacing each line by its Thevenin equivalent, (Fig. 5.20) gives

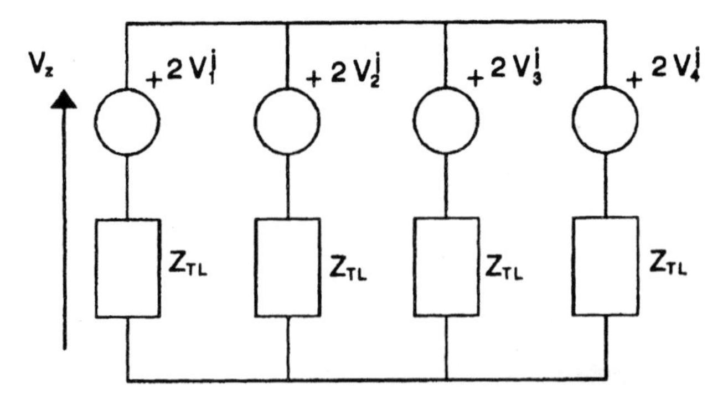

Fig. 5.20 Thevenin equivalent circuit of the shunt node

$$V_z = 0.5\left(V_1^i + V_2^i + V_3^i + V_4^i\right) \tag{5.42}$$

Hence,

$$E_z = -\frac{V_z}{\Delta z} = -0.5\frac{V_1^i + V_2^i + V_3^i + V_4^i}{\Delta z} \tag{5.43}$$

The x-directed current is calculated from the circuit shown in Fig. 5.21 and is

$$I_x = \frac{V_2^i - V_4^i}{Z_{TL}}$$

where Z_{TL} is the characteristic impedance of each transmission line forming the node. It follows that

$$H_y = \frac{V_2^i - V_4^i}{Z_{TL}\Delta y} \tag{5.44}$$

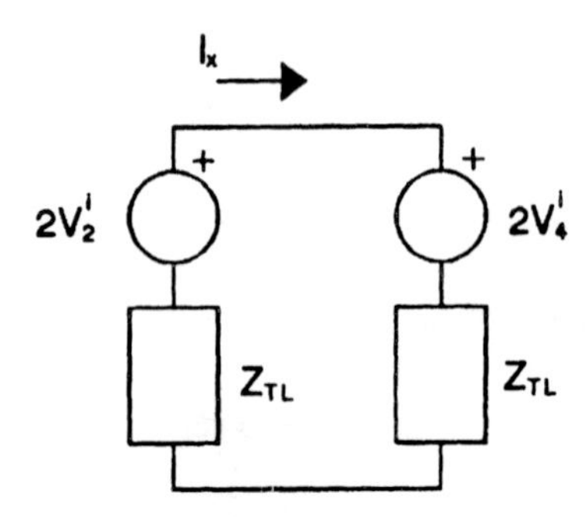

Fig. 5.21 Circuit for calculating the magnetic field component in the y-direction

Similarly,

$$H_x = -\frac{I_y}{\Delta x} = \frac{V_3^i - V_1^i}{Z_{TL}\Delta x} \tag{5.45}$$

The scattered voltages are obtained from (5.20) where now ${}_kV = V_z$. For example, the voltage pulse scattered to port 1 is

$${}_kV_1^r = V_z - {}_kV_1^i = 0.5\left(-{}_kV_1^i + {}_kV_2^i + {}_kV_3^i + {}_kV_4^i\right)$$

The scattering matrix for this node is

$$\mathbf{S} = 0.5\begin{bmatrix} -1 & 1 & 1 & 1 \\ 1 & -1 & 1 & 1 \\ 1 & 1 & -1 & 1 \\ 1 & 1 & 1 & -1 \end{bmatrix} \tag{5.46}$$

The new incident voltages at each node at the next time-step are obtained from the "connection" process, which operates exactly as for the series node [Equation (5.23)]. The logic of the computation is in every way the same as for the series node.

5.2.2 Modeling of Inhomogeneous Lossy Materials

The manner in which inhomogeneous lossy materials are modeled with a shunt node is similar to that described in Section 5.1.2. To avoid repetition, the derivation of the scattering properties of the shunt node with both capacitive and lossy stubs is described directly [10]. The structure is shown in Fig. 5.22a. The link lines 1 to 4 have a characteristic impedance Z_{TL}, and the capacitive and the lossy stubs have characteristic admittance and conductance Y_s and G_s respectively.

It is customary to normalize impedances and admittances using Z_{TL} as the base. Hence, each link line has a normalized impedance of 1, and the stubs have normalized admittances

$$\hat{Y}_s = \frac{Y_s}{Z_{TL}^{-1}},\ \hat{G}_s = \frac{G_s}{Z_{TL}^{-1}}$$

The Thevenin equivalent circuit, with normalized quantities, is shown in Fig. 5.22b. The voltage at the node is then

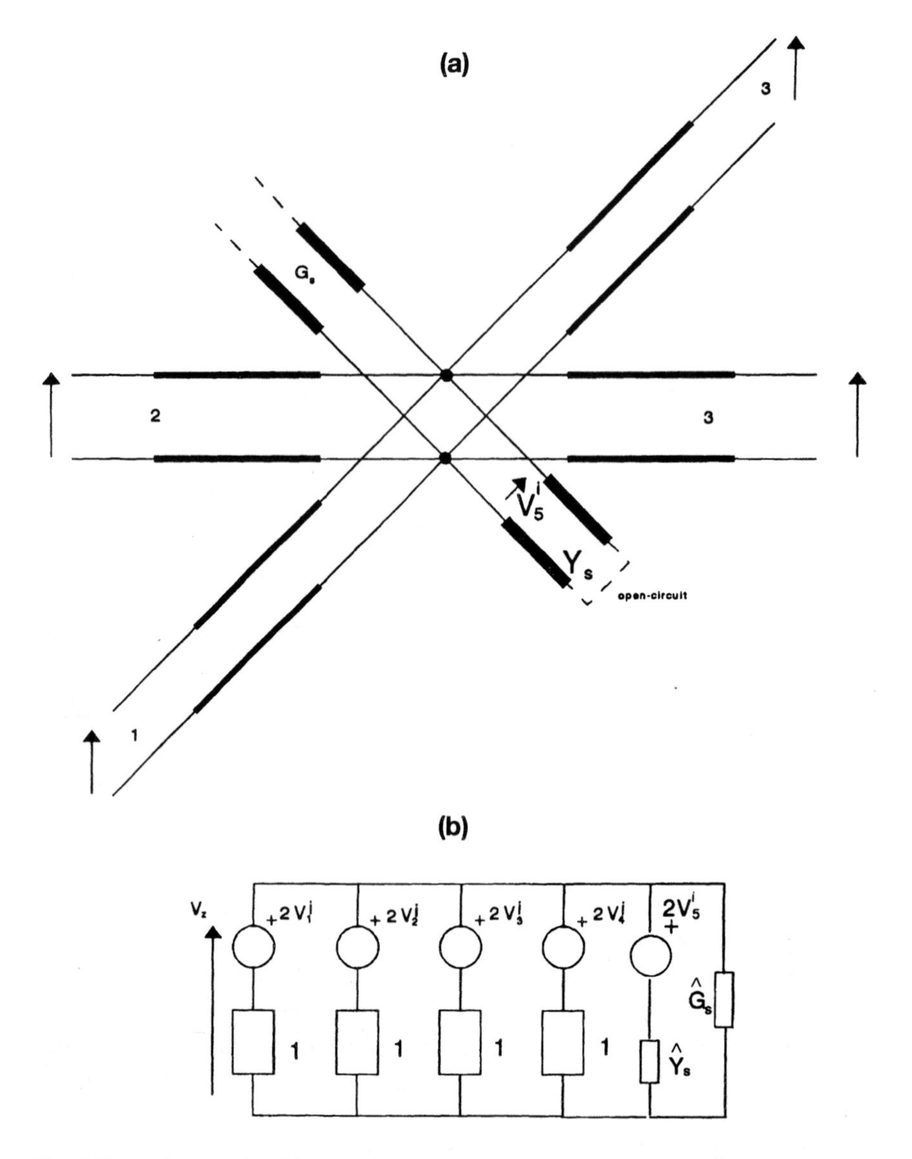

Fig. 5.22 (a) shunt node with capacitive and lossy stubs and (b) its Thevenin equivalent circuit

$$V_z = \frac{2\left(V_1^i + V_2^i + V_3^i + V_4^i\right) + 2V_S^i\,\hat{Y}_S}{4 + \hat{Y}_S + \hat{G}_S} \tag{5.47}$$

and the scattering matrix is

$$\mathbf{S} = \frac{1}{\hat{Y}} \begin{bmatrix} 2-\hat{Y} & 2 & 2 & 2 & 2\hat{Y}_s \\ 2 & 2-\hat{Y} & 2 & 2 & 2\hat{Y}_s \\ 2 & 2 & 2-\hat{Y} & 2 & 2\hat{Y}_s \\ 2 & 2 & 2 & 2-\hat{Y} & 2\hat{Y}_s \\ 2 & 2 & 2 & 2 & 2\hat{Y}_s-\hat{Y} \end{bmatrix} \tag{5.48}$$

where $\hat{Y} = 4 + \hat{Y}_s + \hat{G}_s$. The computation procedure is the same as for lossy series node.

Choice of Parameters. Let us consider a shunt node where each link line has a characteristic impedance Z_{TL}, and where the admittances of the capacitive and lossy stubs are Y_s and G_s, respectively. The parameters modeled can be obtained from Equation (5.29) and are:

Inductance: $Z_{TL}\,\Delta t$
Capacitance: $2\Delta t/Z_{TL} + (\Delta t/2)Y_s$ (effective)
Conductance: G_s

From Kirchhoff's current law

$$\frac{\partial I_x}{\partial x} + \frac{\partial I_y}{\partial y} = -C_d\frac{\partial V_z}{\partial t} - \frac{V_z}{R_d}$$

where C_d and R_d are per unit length quantities; i.e.,

$$C_d = \frac{2\Delta t}{\Delta \ell Z_{TL}}\left(1 + \frac{Y_s Z_{TL}}{4}\right)$$

$$R_d = \frac{\Delta \ell}{G_s}$$

In a lossy medium of conductivity σ, (5.32) becomes

$$\frac{\partial H_x}{\partial y} - \frac{\partial H_y}{\partial x} = -\varepsilon\frac{\partial E_z}{\partial t} - \sigma E_z$$

Hence, the equivalences between model and medium parameters are

$$\mu \leftrightarrow Z_{TL}\frac{\Delta t}{\Delta \ell}, \quad \varepsilon \leftrightarrow \frac{2\Delta t}{\Delta \ell Z_{TL}}\left(1 + \frac{Y_s Z_{TL}}{4}\right), \quad \sigma \leftrightarrow \frac{G_s}{\Delta \ell}$$

EXAMPLE 5.3 *Choice of TLM modeling parameters for a medium of parameters* $(\varepsilon, \mu_0, \sigma)$. The velocity of propagation and the line impedance are chosen to be

$$u_{TL} = \frac{\Delta \ell}{\Delta t} = \sqrt{2}\,\frac{1}{\sqrt{\mu_0 \varepsilon_0}}$$

$$Z_{TL} = \sqrt{2}\sqrt{\frac{\mu_0}{\varepsilon_0}}$$

Then the modeled inductance per unit length is

$$Z_{TL}\frac{\Delta t}{\Delta \ell} = \sqrt{2}\sqrt{\frac{\mu_0}{\varepsilon_0}}\,\frac{\sqrt{\mu_0 \varepsilon_0}}{\sqrt{2}} = \mu_0$$

as desired. The modeled capacitance per unit length is

$$\frac{2\Delta t}{\Delta \ell\, Z_{TL}}\left(1 + Y_s\frac{Z_{TL}}{4}\right) = 2\frac{\sqrt{\mu_0 \varepsilon_0}}{\sqrt{2}\sqrt{2}}\sqrt{\frac{\varepsilon_0}{\mu_0}}\left(1 + Y_s\frac{Z_{TL}}{4}\right) = \varepsilon_0\left(1 + Y_s\frac{Z_{TL}}{4}\right)$$

To model a medium $\varepsilon = \varepsilon_r\,\varepsilon_0$, the capacitive stub admittance must be

$$Y_s = 4\frac{(\varepsilon_r - 1)}{Z_{TL}}$$

Similarly, $G_s = \sigma\,\Delta\ell$.

■

In cases where a complex dielectric constant or the loss tangent is given, the following expressions may be used to relate parameters:

$$\varepsilon^* = \varepsilon\left(1 - j\frac{\sigma}{\omega\varepsilon}\right) = \varepsilon_r\varepsilon_0\,(1 - j\tan\delta)$$

Absorbing boundary conditions in a shunt mesh are imposed by terminating transmission lines by the medium characteristic impedance. Open- and short-circuit boundaries are applied as for the series mesh (Section 5.1.2).

5.3 DISPERSION IN A TWO-DIMENSIONAL MESH

It has already been mentioned that discretization may introduce dispersion. These effects are examined in more detail in this section.

Let us consider propagation at 45° on the mesh shown in Fig. 5.23a. To launch a disturbance at 45° ports 1 and 2 are excited by a 1 V pulse as

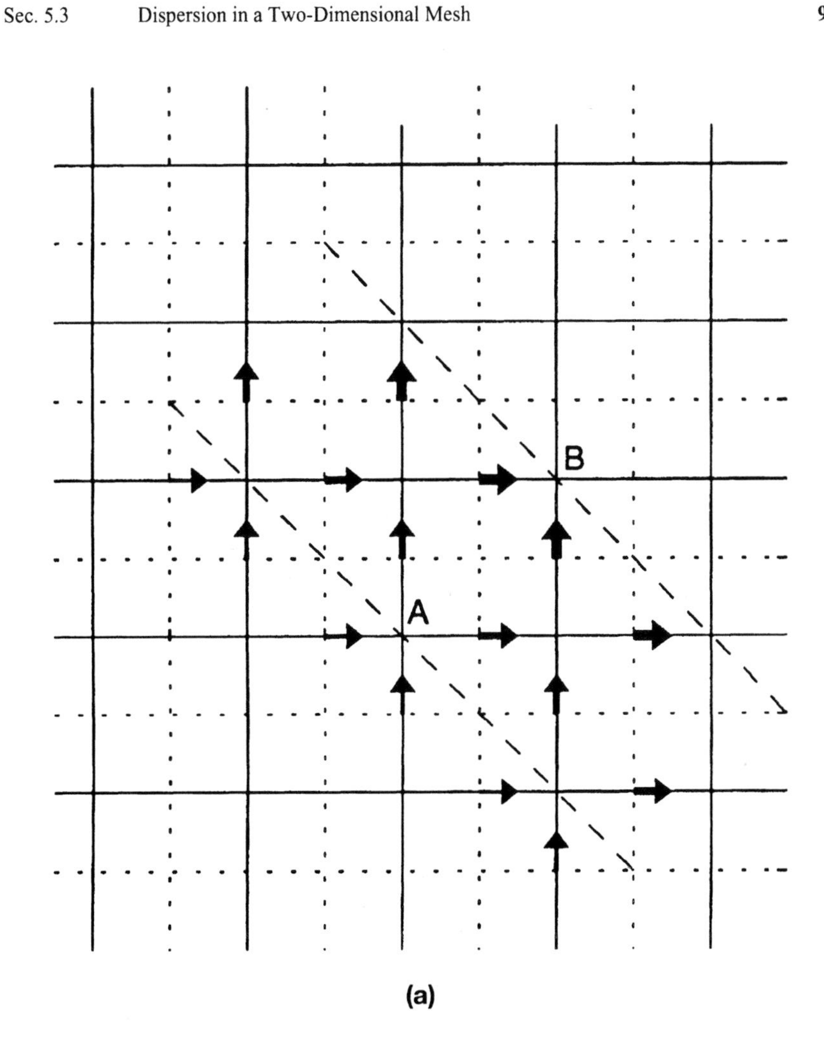

(a)

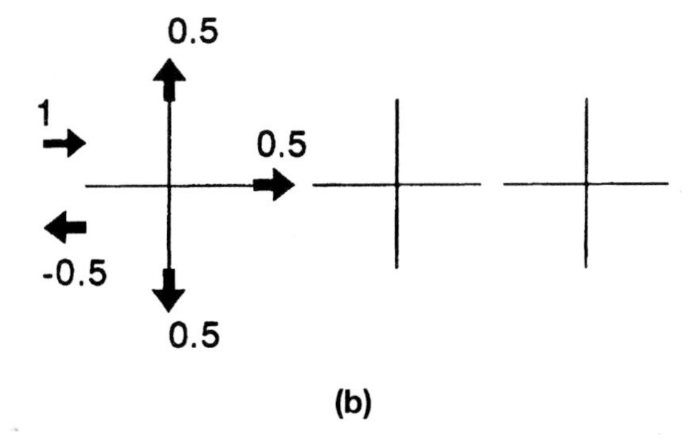

(b)

Fig. 5.23 Schematic of propagation in a 2D mesh; (a) 45° (distance AB is traversed in time $2\Delta t$) and (b) 0°

shown. Using the scattering matrix for the shunt node (5.46), the scattered voltage pulses are found to be

$$V_1^r = 0,\ V_2^r = 0,\ V_3^r = V_4^r = 1\ \text{V}$$

Hence, pulses emerge at ports 3 and 4 unaffected, and the disturbance propagates at 45° without dispersion. The velocity of propagation in the TLM model and at 45° is

$$u_{TLM}^{45^\circ} = \frac{\text{distance traveled}}{\text{time taken}} = \frac{\sqrt{2}\Delta\ell}{2\Delta t} = \frac{u_{TL}}{\sqrt{2}}$$

Next, propagation at 0° along the x-direction is investigated, as shown in Fig 5.23b. With excitation of 1 V applied to port 2 only, the reflected voltages are $V_1^r = V_3^r = V_4^r = 0.5$ V and $V_2^r = -0.5$ V. Therefore, in this case propagation is not ideal, and it is necessary to do a full propagation analysis. Since propagation along the x-direction only is considered, ports 3 and 4 of the shunt node are terminated by an open circuit. Lines 3 and 4 form identical paths in parallel; hence, they can be replaced by a single line of impedance $Z_{TL}/2$ terminated by an open circuit. The structure to be studied is then as shown in Fig. 5.24. Wave propagation on a periodic structure consisting of a cascade of networks of the type shown in Fig. 5.24 is now considered (see also Section 4.3). For lossless propagation, the input and output quantities in each section must be related by the expression

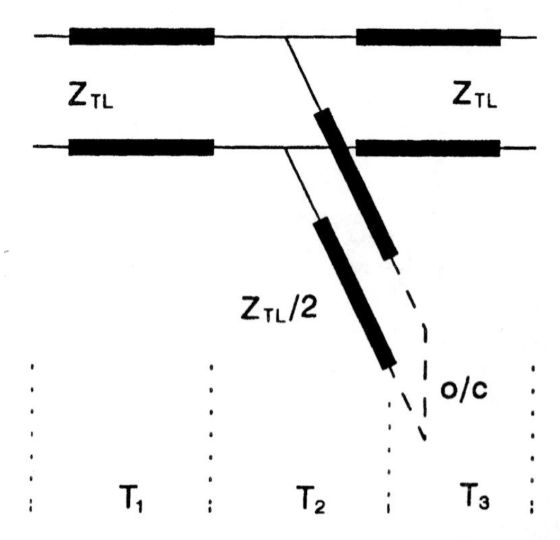

Fig. 5.24 Circuit used to study dispersion for propagation at 0°

$$\begin{bmatrix} V_1 \\ I_1 \end{bmatrix} = \begin{bmatrix} \cos(\beta\ell) & jZ\sin(\beta\ell) \\ \dfrac{j\sin(\beta\ell)}{Z} & \cos(\beta\ell) \end{bmatrix} \begin{bmatrix} V_2 \\ I_2 \end{bmatrix} \tag{5.49}$$

where β is the phase constant.

The network in Fig. 5.24 can be regarded as a cascade of three simpler networks, T_1, T_2, and T_3, as shown. Hence, the overall transmission matrix is:

$$[T] = [T_1][T_2][T_3]$$

$$= \begin{bmatrix} \cos\theta & jZ_{TL}\sin\theta \\ j\dfrac{\sin\theta}{Z_{TL}} & \cos\theta \end{bmatrix} \begin{bmatrix} 1 & 0 \\ j\dfrac{\tan\theta}{Z_{TL}/2} & 1 \end{bmatrix} \begin{bmatrix} \cos\theta & jZ_{TL}\sin\theta \\ j\dfrac{\sin\theta}{Z_{TL}} & \cos\theta \end{bmatrix}$$

where θ is the electrical length of each section ($\theta = \omega\Delta t/2$). Carrying out the multiplication of these matrices and combining with Eq. (5.49) gives

$$\sin(\beta\Delta\ell/2) = \sqrt{2}\sin\left(\frac{\omega\Delta t}{2}\right) \tag{5.50}$$

The free-space wavelength λ_0 is defined as $\lambda_0 = u_{TL}/f$, where u_{TL} is the propagation velocity on each line. Hence,

$$\frac{\omega\Delta t}{2} = \frac{\omega}{2}\frac{\Delta\ell}{u_{TL}} = \frac{2\pi f}{2}\frac{\Delta\ell}{u_{TL}} = \pi\frac{u_{TL}}{\lambda_0}\frac{\Delta\ell}{u_{TL}} = \pi\frac{\Delta\ell}{\lambda_0} \tag{5.51}$$

The wave propagation velocity in the model is

$$u_{TLM} = \frac{\omega}{\beta} = \frac{2\pi f}{\beta} = \frac{2\pi}{\beta}\frac{u_{TL}}{\lambda_0}, \text{ hence}$$

$$\beta = \frac{2\pi}{\lambda_0}\frac{u_{TL}}{u_{TLM}} \tag{5.52}$$

Substituting $\omega\Delta t/2$ and β from Equations (5.51) and (5.52) into Equation (5.50) gives

$$\frac{u_{TLM}}{u_{TL}} = \frac{\pi\left(\dfrac{\Delta\ell}{\lambda_0}\right)}{\sin^{-1}\left[\sqrt{2}\sin\left(\pi\dfrac{\Delta\ell}{\lambda_0}\right)\right]} \tag{5.53}$$

This expression relates the wave propagation velocity in the model to the propagation velocity on each line. Clearly, the relationship is dependent on the ratio of $\Delta\ell$ to the wavelength. A limiting case exists when $\Delta\ell/\lambda_0 = 1/4$ corresponding to the first network cutoff frequency. At this limiting value,

$$\frac{u_{TLM}}{u_{TL}} = 0.5$$

To model a medium with wave velocity u_w the propagation velocity is chosen to be $u_{TL} = \sqrt{2}u_w$ (see example in Section 5.2.2). Hence,

$$\frac{u_{TLM}}{u_w} = \frac{1}{\sqrt{2}}$$

At the low-frequency limit ($\Delta\ell/\lambda_o$<<), $\sin(\pi\Delta\ell/\lambda_o) \cong \pi\Delta\ell/\lambda_o$, hence

$$\frac{u_{TLM}}{u_{TL}} \to \frac{1}{\sqrt{2}}, \text{ therefore } \frac{u_{TLM}}{u_w} \to 1$$

At a frequency where $\Delta\ell/\lambda_0 = 1/10$,

$$\frac{u_{TLM}}{u_w} = \sqrt{2}\ 0.694 = 0.981$$

In this case, the model wave velocity deviates by less than 2 percent from its desired value. The velocity characteristics for the shunt node where u_{TL} is chosen to be equal to $\sqrt{2}u_w$ are shown in Fig. 5.25. Curves are shown for propagation at 45° and 0°. Velocity errors lie between these two curves. A suitable choice of $\Delta\ell$ to give an acceptable maximum velocity error at the highest frequency of interest can be made using these curves.

A more detailed study of dispersion in a general TLM mesh may be found in Ref. [12]. A better view of propagation in a 2D TLM mesh may be obtained by plotting the wave number vector normalized to the medium wave number as shown in Fig. 5.26a for the ideal case of a very fine discretization ($\Delta\ell/\lambda_0 \to 0$) and for two finite values of discretization $\Delta\ell/\lambda_0 = 0.1$ and 0.2. The percentage error for different angles of propagation is shown more clearly in Fig. 5.26b.

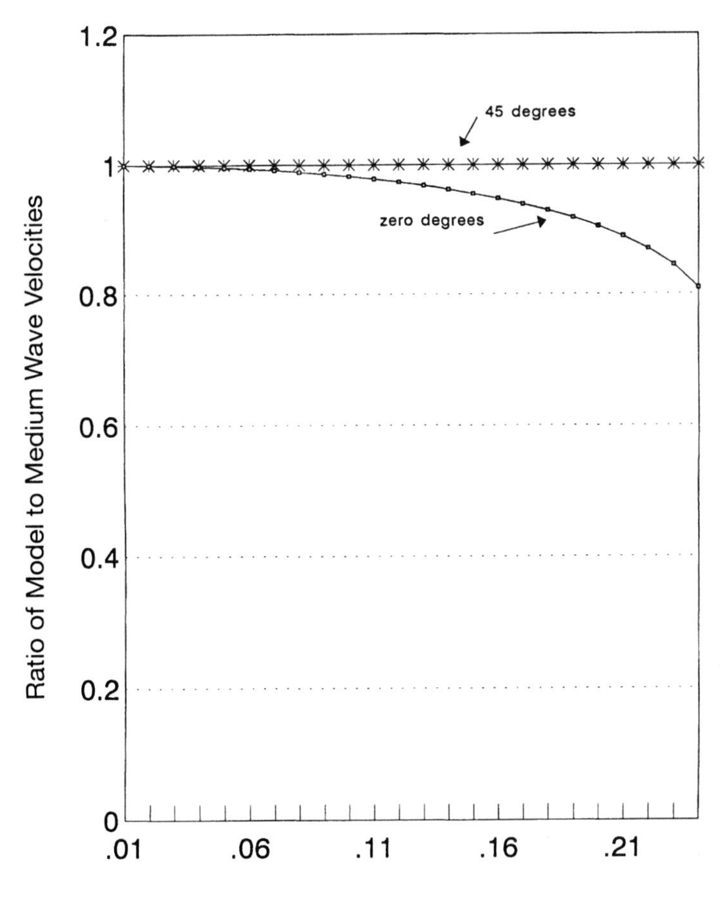

Fig. 5.25 Dispersion effects in a 2D mesh

5.4 DUALITY IN ELECTROMAGNETICS

It is possible by applying the principle of duality [11] to use either the series or the shunt nodes to model both TE and TM fields by establishing a different analogy between circuit and field quantities. This is explained more fully in [4, 5]. The difficulties which may arise when using dual networks are discussed in detail in section 3.4.7 of [3]. Most users will feel more comfortable avoiding duality and using either the series or the shunt node depending on the nature of the problem. Both nodes were necessary during the development of TLM to solve three-dimensional problems, as is discussed in the next chapter.

(a)

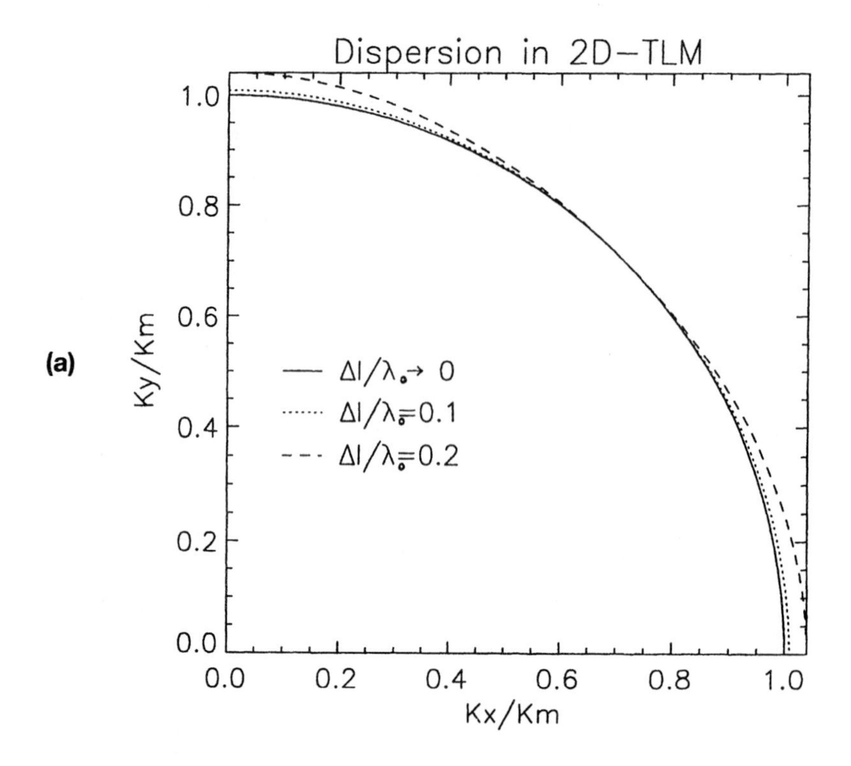

(b)

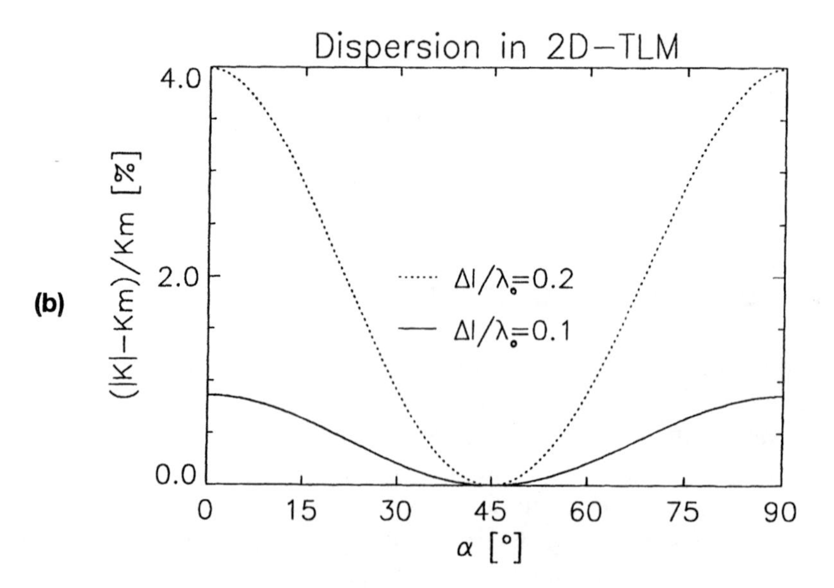

Fig. 5.26 (a) wave vector (*kx*, *ky*) for different levels of discretization and (b) percentage error due to dispersion for different angles of propagation

REFERENCES

[1] Huygens, C. 1690. Traite de la Lumiere. Paris: Leiden.

[2] Johns, P.B. 1974. A new mathematical model to describe the physics of propagation. *Radio and Electronic Engineer* 44, 657–666.

[3] Hoefer, W.J.R., and P.P.M. So. 1991. *The Electromagnetic Wave Simulator.* New York: John Wiley & Sons.

[4] Akhtarzad, S., and P.B. Johns. 1975. Generalised elements for TLM method of numerical analysis. *Proceedings of IEE* 122, 1349–1352.

[5] Johns, P.B. 1982. Ideal transformers and gauge transformations in lumped network models of electromagnetic fields. *Proceedings of IEE* 129-A, 381–386.

[6] Naylor, P. 1991. Utilising 2-D series TLM for modelling TE nodes in dielectric and lossy environments. *Electronics Letters* 27, 1499–1501.

[7] Simons, N.R.S., and E. Bridges, E. 1990. Method for modelling free space boundaries in TLM situations. *Electronics Letters* 26, 453–455.

[8] Hoefer, W.J.R. 1989. The discrete time domain Green's Function or Johns Matrix—A new powerful concept in TLM. *International Journal of Numerical Modelling* 2, 215–225.

[9] Johns, P.B., and R.L. Beurle. 1971. Numerical solution of 2-dimensional scattering problems using a transmission-line matrix," *Proceedings of IEE* 118, 1203–1208.

[10] Johns, P.B. 1974. The solution of inhomogeneous waveguide problems using a transmission-line matrix. *IEEE Transactions* MTT-22, 209–215.

[11] Whinnery, J.R., and S. Ramo. 1944. A new approach to the solution of high-frequency field problems. *Proceedings of IRE* 32, 284–288.

[12] Nielsen, J.S. 1992. "TLM Analysis of Microwave and Millimeter-Wave Structures with Embedded Nonlinear Devices." PhD Thesis, University of Ottawa.

6

Three-Dimensional TLM Models

Most problems encountered in engineering applications are three-dimensional, and it is therefore essential to develop models to describe 3D fields. The development of 3D TLM nodes has followed several stages which are briefly described in the next section. There follows a more extensive coverage of the symmetrical condensed node, which is now the most commonly used structure for TLM simulation. More powerful techniques, such as the hybrid node and the multigrid TLM mesh, are then described.

6.1 THE DEVELOPMENT OF THREE-DIMENSIONAL NODES

In the preceding chapter, the series and shunt nodes were described, and it was shown that each describes three electromagnetic field components in both TE and TM modes. Intuitively, it therefore appears possible that a combination of series and shunt nodes may be used to model all six electromagnetic components in three-dimensional space [1].

Let us consider a shunt node on the x-z plane connected to series nodes on the y-z and y-x planes as shown in Fig. 6.1. Assuming that voltages and currents represent electric and magnetic fields, respectively, Kirchhoff's current law (KCL) in the shunt node describes the following equation:

$$\frac{\partial H_x}{\partial z} - \frac{\partial H_z}{\partial x} = \varepsilon \frac{\partial E_y}{\partial t} \tag{6.1}$$

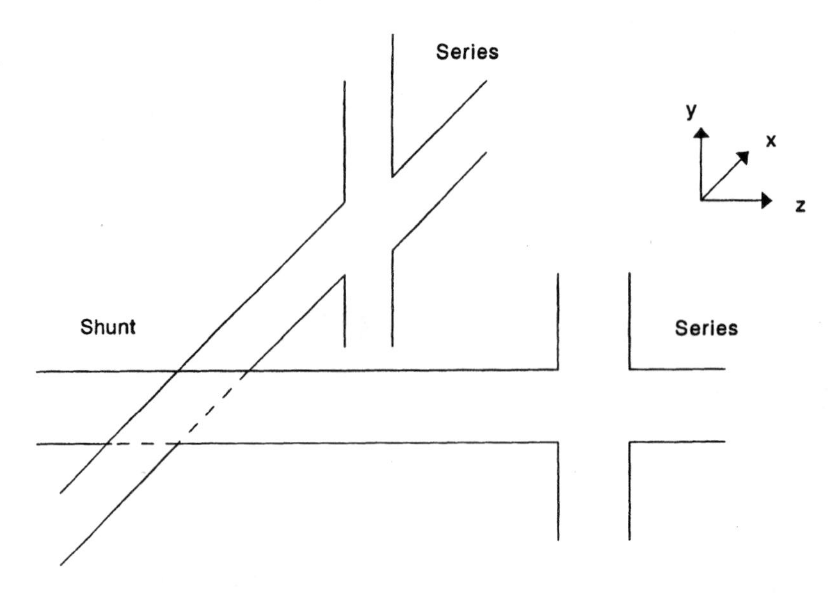

Fig. 6.1 Part of a 3D node containing two series and one shunt node

Similarly, from KVL for the series node on the *y-z* plane,

$$\frac{\partial E_z}{\partial y} - \frac{\partial E_y}{\partial z} = -\mu \frac{\partial H_x}{\partial t} \tag{6.2}$$

and from the series node on the *x-y* plane,

$$\frac{\partial E_y}{\partial x} - \frac{\partial E_x}{\partial y} = -\mu \frac{\partial H_z}{\partial t} \tag{6.3}$$

Equations (6.1) through (6.3) are a subset of Maxwell's equations. The remaining equations may be described by a series node on the *x-z* plane connected to shunt nodes in the *y-z* and *x-y* planes. Connecting these two circuits gives the arrangement shown in Fig. 6.2. This structure is known as the *expanded node,* the name deriving from the time delay of half a time-step between series and shunt nodes. Further details about the expanded node may be found in Ref. [1].

The disadvantages of this structure lie in its complexity. Using the normal analogy between circuit and field quantities (described in the last chapter) leads to the calculation of different field components and polarizations at points that are physically separated. This causes difficulties in applying boundary conditions simply and correctly. The manner in which lossy and inhomogeneous materials may be modeled was described in Refs. [2] and [3], and the implementation of an irregularly spaced or

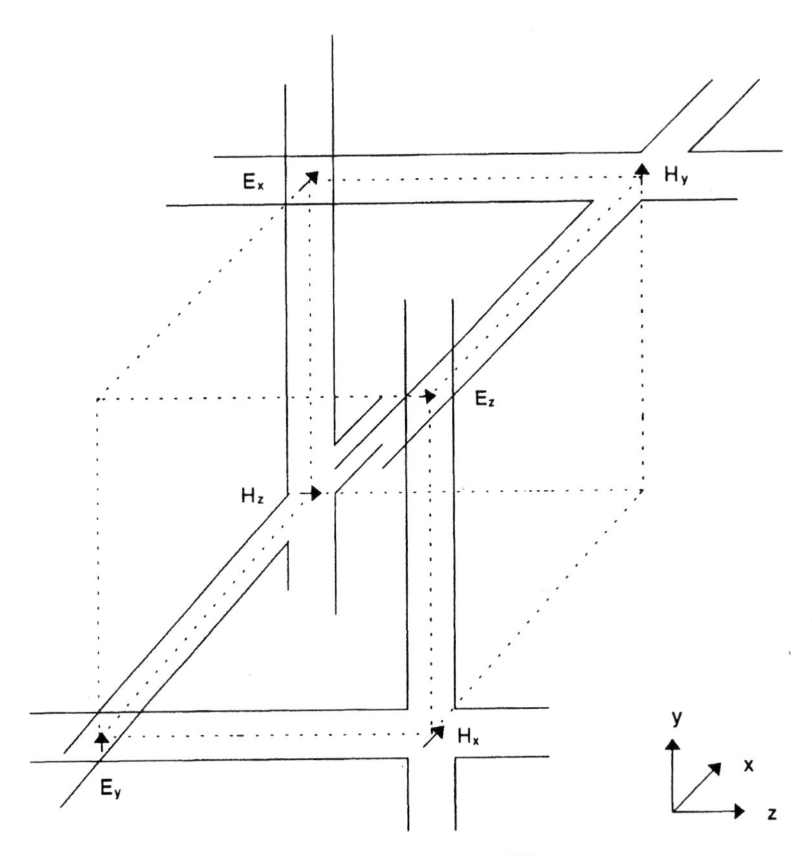

Fig. 6.2 A three-dimensional TLM node

"graded" TLM mesh was presented in Refs. [4] and [5]. With regard to stubs and terminations, the same principles as described in connection with 2D nodes also apply for this node. The expanded node has been used extensively and successfully for many years to study a variety of electromagnetic problems [6].

A further development of the 3D node was described in Refs. [7] and [8]. This node, referred to as the *punctual* or *asymmetrical condensed node,* has the advantage of modeling all field components at the same point. It is, however, asymmetrical, since the first connection to the node is either to a shunt or a series node, depending on direction. Further discussion of the advantages and disadvantages of these nodes may be found in Refs. [6] and [9].

A structure that aims to combine the best features and minimize the disadvantages of the expanded and asymmetrical condensed nodes was first described by Johns in 1986 [10] and is shown in Fig. 6.3. It consists of 12 ports to represent 2 polarizations in each coordinate direction. The

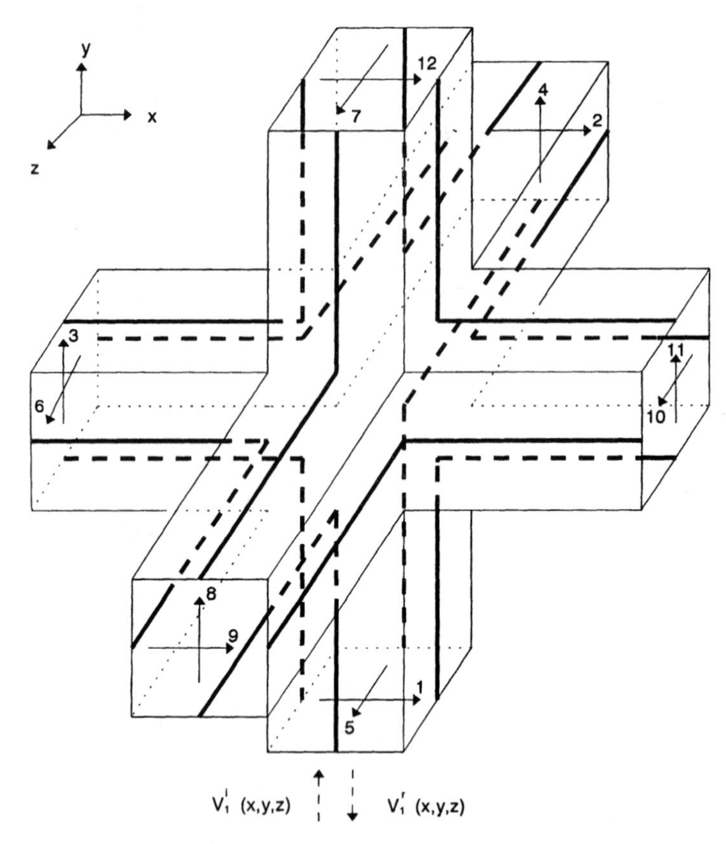

Fig. 6.3 The three-dimensional symmetrical condensed node (SCN). Node coordinates are (x, y, z). The incident and reflected voltages for port 1 are indicated.

voltage pulses corresponding to the 2 polarizations are carried on transmission line pairs shown schematically in heavy line. The two lines such as 8 and 9, which in the expanded node are separated in space, do not directly couple with each other. This node, which is referred to as the *symmetrical condensed node* (SCN), is described in detail in the next section.

6.2 THE SYMMETRICAL CONDENSED NODE

The topology of the SCN shown in Fig. 6.3 does not lend itself to a treatment using Thevenin equivalent circuits. Instead, the scattering properties are obtained from general energy and charge conservation principles as described in Ref. [9]. The node may be viewed as the bringing together of the three structures shown in Fig. 6.4, each representing lines in the three coordinate planes. The scattering matrix S relating reflected V^r to incident V^i voltages is a 12×12 matrix. Any pulse incident onto a port can, in gen-

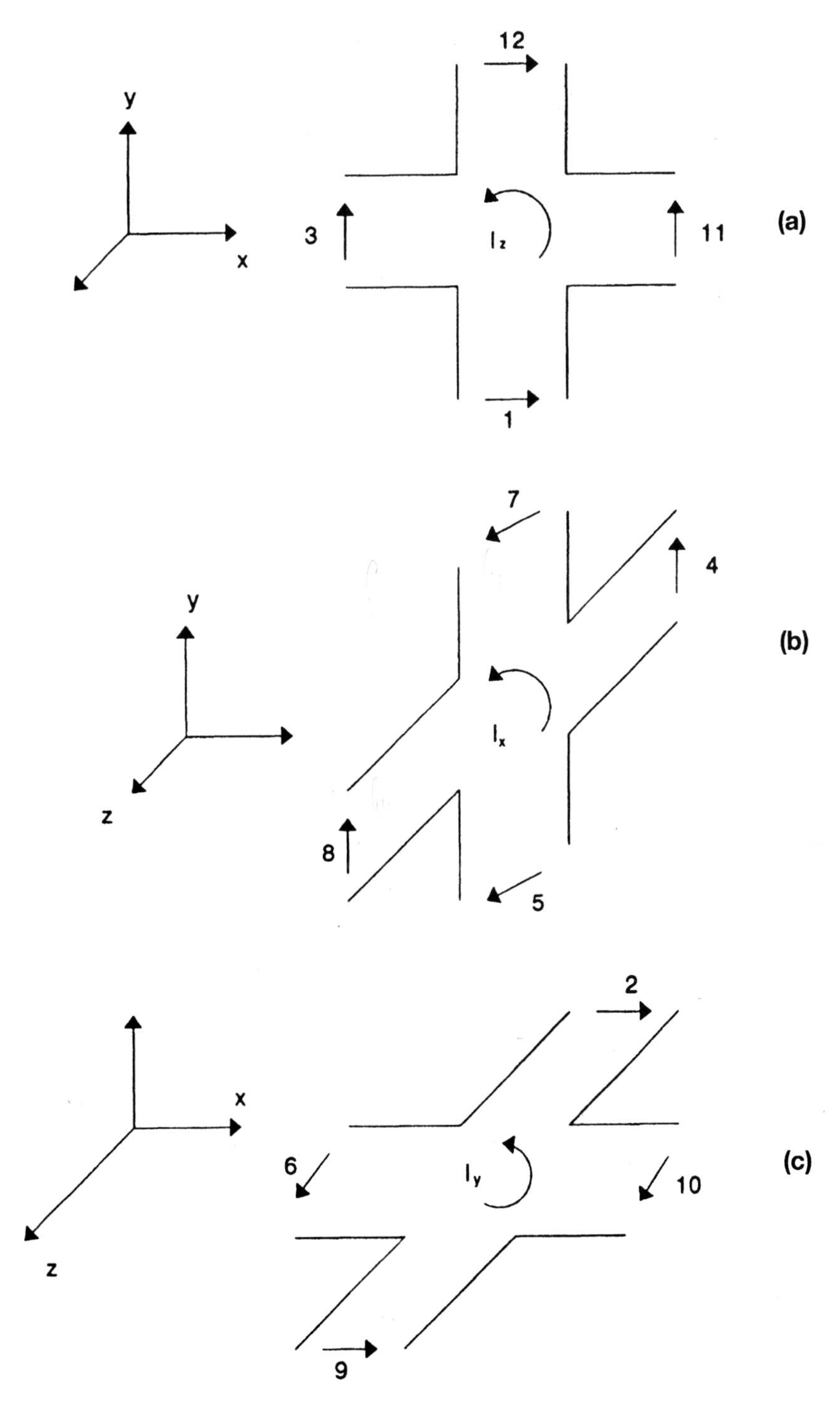

Fig. 6.4 Three transmission-line clusters (part of the SCN)

eral, couple to all other ports, so the first task is to ascertain which coupling paths are possible and to establish an analogy between circuit and field quantities. The elements of the scattering matrix will then be calculated.

Let us consider a voltage pulse incident on port 1 of the SCN. Since this pulse is x-directed, it is associated with E_x and, since it contributes to current I_z as shown in Fig. 6.4a, it is also associated with H_z. Maxwell's equations relating E_x and H_z are

$$\frac{\partial H_z}{\partial y} - \frac{\partial H_y}{\partial z} = \varepsilon \frac{\partial E_x}{\partial t} \tag{6.4}$$

$$\frac{\partial E_y}{\partial x} - \frac{\partial E_x}{\partial y} = -\mu \frac{\partial H_z}{\partial t} \tag{6.5}$$

Equation (6.4) implies that V_1, in principle, can couple into port 1 (i.e., reflected), ports 2 and 9 (since they are associated with E_x and H_y, and port 12 (associated with E_x and H_z). Similarly, Equation (6.5) implies that, in addition, it must couple to ports 3 and 11, since they are associated with E_y and H_z.

A more intuitive approach may be adopted in establishing this coupling, as shown in Fig. 6.5. Let the incident pulse on port 1 be equal to one volt. Then, an amount (a) may be reflected, and an amount (b) may be coupled, to ports 2 and 9 as shown. Symmetry dictates that 2 and 9 couple identically to port 1. Similarly, an amount (c) will propagate directly across the junction to couple into port 12. The field vector may rotate as it encounters the junction to couple into port 3 (d) and port 11 (–d). The minus sign is due to the polarity of the pulse coupling to port 11. The elements of the first column of the scattering matrix have now been identified. This procedure may be applied to all ports to identify all non-zero elements. The overall matrix is shown below.

$$\mathbf{S} = \begin{bmatrix}
a & b & d & 0 & 0 & 0 & 0 & 0 & b & 0 & -d & c \\
b & a & 0 & 0 & 0 & d & 0 & 0 & c & -d & 0 & b \\
d & 0 & a & b & 0 & 0 & 0 & b & 0 & 0 & c & -d \\
0 & 0 & b & a & d & 0 & -d & c & 0 & 0 & b & 0 \\
0 & 0 & 0 & d & a & b & c & -d & 0 & b & 0 & 0 \\
0 & d & 0 & 0 & b & a & b & 0 & -d & c & 0 & 0 \\
0 & 0 & 0 & -d & c & b & a & d & 0 & b & 0 & 0 \\
0 & 0 & b & c & -d & 0 & d & a & 0 & 0 & b & 0 \\
b & c & 0 & 0 & 0 & -d & 0 & 0 & a & d & 0 & b \\
0 & -d & 0 & 0 & b & c & b & 0 & d & a & 0 & 0 \\
-d & 0 & c & b & 0 & 0 & 0 & b & 0 & 0 & a & d \\
c & b & -d & 0 & 0 & 0 & 0 & 0 & b & 0 & d & a
\end{bmatrix} \tag{6.6}$$

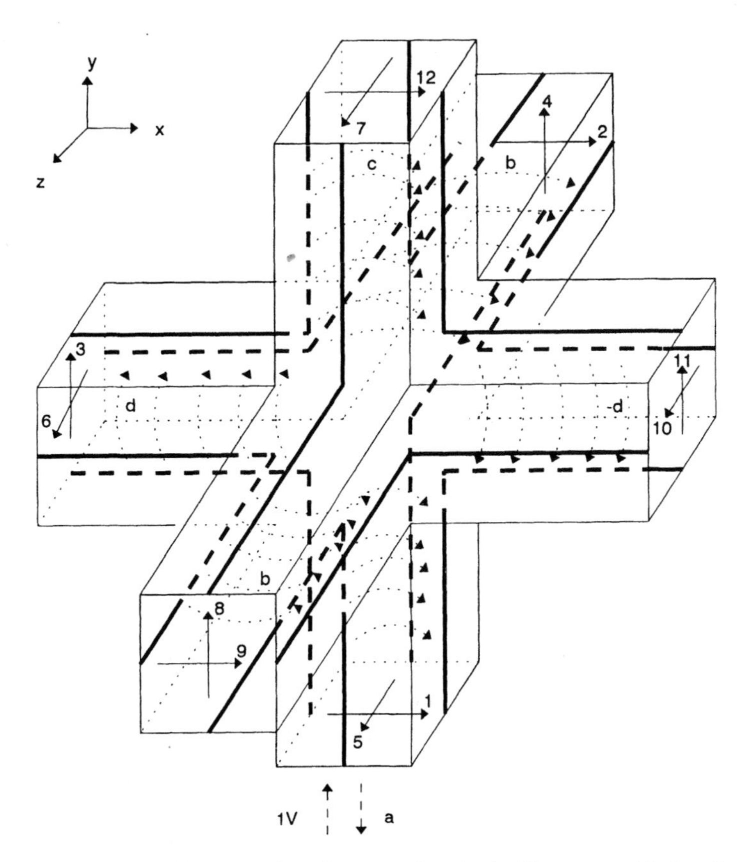

Fig. 6.5 Schematic of the scattering of a one-volt pulse incident on port 1. A voltage pulse (a) is reflected back, and the amounts indicated are scattered into ports 2, 3, 9, 11, and 12.

It now remains to determine the value of each of the unknown coupling parameters a, b, c, and d. These will be determined by demanding that, for a lossless network, the total incident power must be equal to the total reflected power. This results in the unitary condition for the scattering matrix namely that [11]

$$\mathbf{S}^{T}\mathbf{S} = \mathbf{I} \tag{6.7}$$

where the superscript T stands for the transpose matrix, and **I** is the unit matrix. Imposing this condition on matrix (6.6) gives the following system of equations:

$$a^2 + 2b^2 + 2d^2 + c^2 = 1 \qquad 2b(a+c) = 0$$

$$2d(a-c) = 0 \qquad 2ac + 2b^2 - 2d^2 = 0 \tag{6.8}$$

There are several possible solutions to these equations. The one chosen must also satisfy additional constraints determined by Kirchhoff's Laws or the analogous Maxwell's equations.

Following the same approach as for the 2D nodes the following equivalence between field and circuit parameters is established.

$$
\begin{aligned}
E_x &= -V_x / \Delta x \\
E_y &= -V_y / \Delta y \\
E_z &= -V_z / \Delta z \\
H_x &= I_x / \Delta x \\
H_y &= I_y / \Delta x \\
H_z &= I_z / \Delta x
\end{aligned}
\tag{6.9}
$$

where

Δx, Δy, Δz = dimensions of the block of space described by the node

V_x, V_y, V_z = the x-, y-, and z-directed voltages across the corresponding ports

I_x, I_y, I_z = the circulating currents shown in Fig. 6.4

Designating the current entering port n by I_n, it is then possible to evaluate each term in Equation (6.4) in terms of circuit quantities.

$$\frac{\partial H_z}{\partial y} = \frac{H_z(y+\Delta y) - H_z(y)}{\Delta y} = \frac{(-I_{12}) - (I_1)}{\Delta z \Delta y}$$

$$\frac{\partial H_y}{\partial z} = \frac{H_y(z+\Delta z) - H_y(z)}{\Delta z} = \frac{I_g - (-I_2)}{\Delta y \Delta z}$$

$$\frac{\partial E_x}{\partial t} = -\frac{1}{\Delta z}\frac{\partial V_z}{\partial t}$$

and substituting in Equation (6.4), where

$$C_x = \varepsilon \frac{\Delta y \Delta z}{\Delta x}$$

(which is the total capacitance associated with lines 1, 2, 9, and 12), gives

$$I_1 + I_2 + I_9 + I_{12} = C_x \frac{\partial V_x}{\partial t} \tag{6.10}$$

This is a continuity equation which states that there can be no loss of charge at the node. Assuming that all lines have the same characteristic impedance Z and that one volt is applied to V_1, using the parameters of the first column of the scattering matrix in Equation (6.6) gives

$$I_1 = \frac{(1-a)}{Z}, I_2 = \frac{-b}{Z}, I_9 = \frac{-b}{Z}, I_{12} = \frac{-c}{Z}$$

Substituting in Equation (6.10) gives

$$(1-a) = 2b + c \tag{6.11}$$

An alternative derivation of this formula, based on charge conservation, is to demand that the total charge entering port 1 must be equal to the total charge emerging from all the other ports coupled to port 1 (first column of scattering matrix).

The terms in Equation (6.5) can also be evaluated to give its discrete circuit equivalent.

$$\frac{-E_y(x+\Delta x) - (-E_y(x))}{\Delta x} - \frac{-E_x(y+\Delta y) - (-E_x(y))}{\Delta y} = -\frac{L_z \Delta z}{\Delta x \Delta y} \frac{\partial H_z}{\partial t}$$

where $L_z = \mu \frac{\Delta x \Delta y}{\Delta z}$ is the inductance associated with lines 1, 3, 11, 12.

Substituting in terms of circuit parameters using Equation (6.9) gives

$$-V_1 + V_3 - V_{11} + V_{12} = L_z \frac{\partial I_z}{\partial t} \tag{6.12}$$

This equation expresses KVL in the mesh shown in Fig. 6.4a. Assuming as before that one volt is applied on port 1, and substituting in terms of the scattering parameters, gives

$$-(1+a) + d - (-d) + c = 0$$

or

$$1 + a = 2d + c \tag{6.13}$$

The solution of Equations (6.8), which also satisfies (6.11) and (6.13), is then easily found, and is

$$a = 0, b = 0.5, c = 0, d = 0.5$$

The complete scattering matrix is shown below and is particularly simple—the voltage scattered to each port is the result of a simple arithmetic operation on four incident voltages.

$$\mathbf{S} = 0.5\begin{bmatrix} 0 & 1 & 1 & 0 & 0 & 0 & 0 & 0 & 1 & 0 & -1 & 0 \\ 1 & 0 & 0 & 0 & 0 & 1 & 0 & 0 & 0 & -1 & 0 & 1 \\ 1 & 0 & 0 & 1 & 0 & 0 & 0 & 1 & 0 & 0 & 0 & -1 \\ 0 & 0 & 1 & 0 & 1 & 0 & -1 & 0 & 0 & 0 & 1 & 0 \\ 0 & 0 & 0 & 1 & 0 & 1 & 0 & -1 & 0 & 1 & 0 & 0 \\ 0 & 1 & 0 & 0 & 1 & 0 & 1 & 0 & -1 & 0 & 0 & 0 \\ 0 & 0 & 0 & -1 & 0 & 1 & 0 & 1 & 0 & 1 & 0 & 0 \\ 0 & 0 & 1 & 0 & -1 & 0 & 1 & 0 & 0 & 0 & 1 & 0 \\ 1 & 0 & 0 & 0 & 0 & -1 & 0 & 0 & 0 & 1 & 0 & 1 \\ 0 & -1 & 0 & 0 & 1 & 0 & 1 & 0 & 1 & 0 & 0 & 0 \\ -1 & 0 & 0 & 1 & 0 & 0 & 0 & 1 & 0 & 0 & 0 & 1 \\ 0 & 1 & -1 & 0 & 0 & 0 & 0 & 0 & 1 & 0 & 1 & 0 \end{bmatrix} \quad (6.14)$$

To achieve synchronism, the block of space represented by the SCN is chosen to be a cube $\Delta x = \Delta y = \Delta z = \Delta \ell$, thus producing what is described as a *regular mesh*. This has limitations as regards modeling of nonuniformities and fine features, which will be tackled in Section 6.3. Before generalizing the SCN by adding stubs, some practical issues will be explored. These can be most simply understood for the regular mesh just described.

6.2.1 Propagation Properties in a Regular SCN Mesh

Let us consider propagation in the x-direction of a plane wave with the electric field polarized in the y-direction. This requires the excitation of port 3 on all the nodes on a y-z plane. A cluster of such nodes is shown schematically in Fig. 6.6. Exciting port 3 only on each of these nodes ${}_kV_3^i = 1$ will produce the following scattered pulses

$${}_kV_1^r = {}_kV_4^r = {}_kV_8^r = -{}_kV_{12}^r = 0.5$$

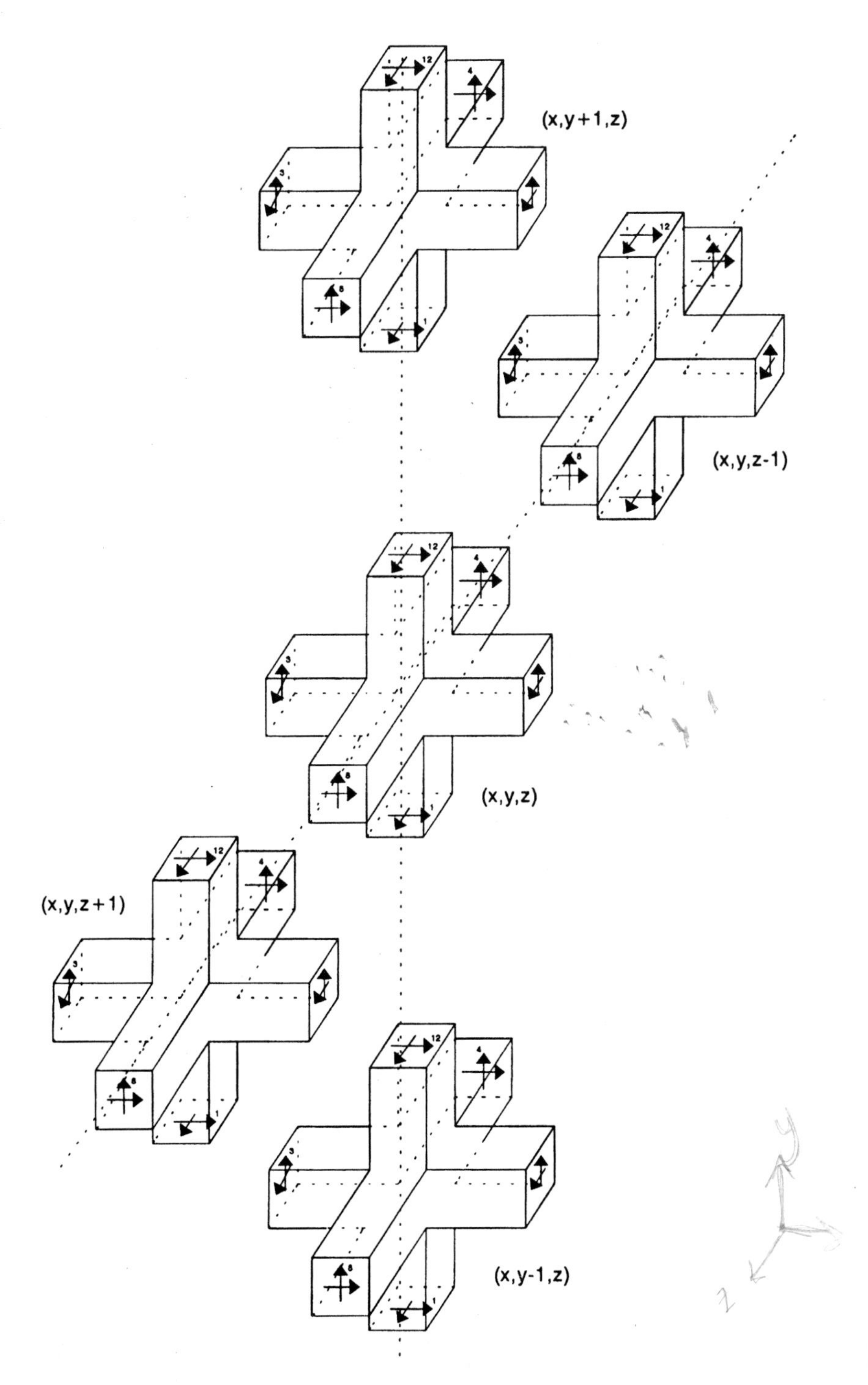

Fig. 6.6 A cluster of SCN nodes on the *y-z* plane

All other scattered pulses are zero, as can be easily confirmed from the scattering matrix. This means that no energy is reflected back ($_kV_1^r = 0$), and no energy is transmitted through ($_kV_{10}^r = {_kV_{11}^r} = 0$). At the next time-step, the incident voltages on node (x, y, z) will be coming from neighboring nodes. Port 1 of (x, y, z) will receive the pulse reflected into port 12 of node (x, y – 1, z) at the previous time step; i.e.,

$$_{k+1}V_1^i(x, y, z) = {_kV_{12}^r}(x, y-1, z) = -0.5$$

Similarly,

$$_{k+1}V_4^i(x, y, z) = {_kV_8^r}(x, y, z-1) = 0.5$$

$$_{k+1}V_8^i(x, y, z) = {_kV_4^r}(x, y, z+1) = 0.5$$

$$_{k+1}V_{12}^i(x, y, z) = {_kV_1^r}(x, y+1, z) = 0.5$$

All other incident voltages on node (x, y, z) at the start of time-step $k + 1$ are zero. The reflected voltages into each port of node (x, y, z) can then be readily obtained from **S** and are all zero except for $_{k+1}V_{11}^r$, which is equal to 1. Hence, the wave propagates without dispersion, and it takes two time-steps to cover the distance $\Delta\ell$. The choice of space and time discretization must then be such that

$$\frac{\Delta l}{\Delta t} = 2u \tag{6.15}$$

where u is the medium propagation velocity. Propagation at 45° may also be studied by exciting ports 3 and 4 simultaneously with all other incident voltages set to zero. From the scattering matrix the reflected voltages are found to be

$$V_1^r = V_3^r = V_{11}^r = -V_{12}^r = 0.5$$

and

$$V_4^r = V_5^r = -V_7^r = -V_8^r = 0.5$$

The first equation above shows scattering equivalent to that on a 2D series node such as the one shown in Fig. 6.4a, while the second equation represents scattering on the node shown in Fig. 6.4b. The propagation properties of this node were studied in Section 5.1, and it was found that the velocity of propagation in the model u_{TLM} is equal to $u_{TL}/\sqrt{2}$, where u_{TL} is the velocity of propagation on each line. The distance traveled for propagation at 45° in the 3D problem is $\sqrt{2}\ \Delta\ell/2$; hence,

$$u_{TLM} = \frac{u_{TL}}{\sqrt{2}} = \frac{\sqrt{2}\Delta\ell}{2\Delta t}\frac{1}{\sqrt{2}} = \frac{\Delta\ell}{2\Delta t}$$

This indicates that the choice of parameters in Equation (6.15) will give the correct low-frequency propagation velocity at 45°. Further discussion of the dispersion properties of the SCN may be found in Refs. [12–14].

6.2.2 Computation in an SCN Mesh

Computation in 3D follows closely the procedures described in connection with 2D meshes. The calculation starts by imposing the initial conditions and excitation. Excitation is described in detail in Section 6.2.4. This process defines all incident voltages in the mesh at the start of the simulation ($k = 0$).

The next step is "scattering," whereby the reflected voltages at each node are obtained from

$$\mathbf{V}^r = \mathbf{S}\ \mathbf{V}^i \tag{6.16}$$

The scattering matrix **S** has the form given by Equation (6.14) for nodes describing a medium of parameters (ε, μ). It also is possible to insert conducting bodies in the simulation space. One way to represent a part of space as a perfect conductor is to describe it by the short-circuit node shown schematically in Fig. 6.7. Many such nodes may be joined together to form conducting wire structures. Clearly, in a short-circuit node the reflected voltage into each port is simply equal to minus the incident voltage; i.e.,

$$V_n^r = -V_n^i$$

The scattering matrix for the short-circuit node is diagonal with elements equal to −1. Other forms of **S** are possible to describe nonuniform, lossy, materials, as will be discussed in subsequent sections. However, scattering in all cases proceeds according to Equation (6.16) with the matrix **S** to suit the properties of the medium being modeled.

From the scattered voltages, it is straightforward to obtain the incident voltages at the next time-step on all ports and nodes in the problem. This procedure is similar to that used in 2D simulation, and it is described as "connection." Clearly, the incident voltage on port 1 of node (x, y, z) at

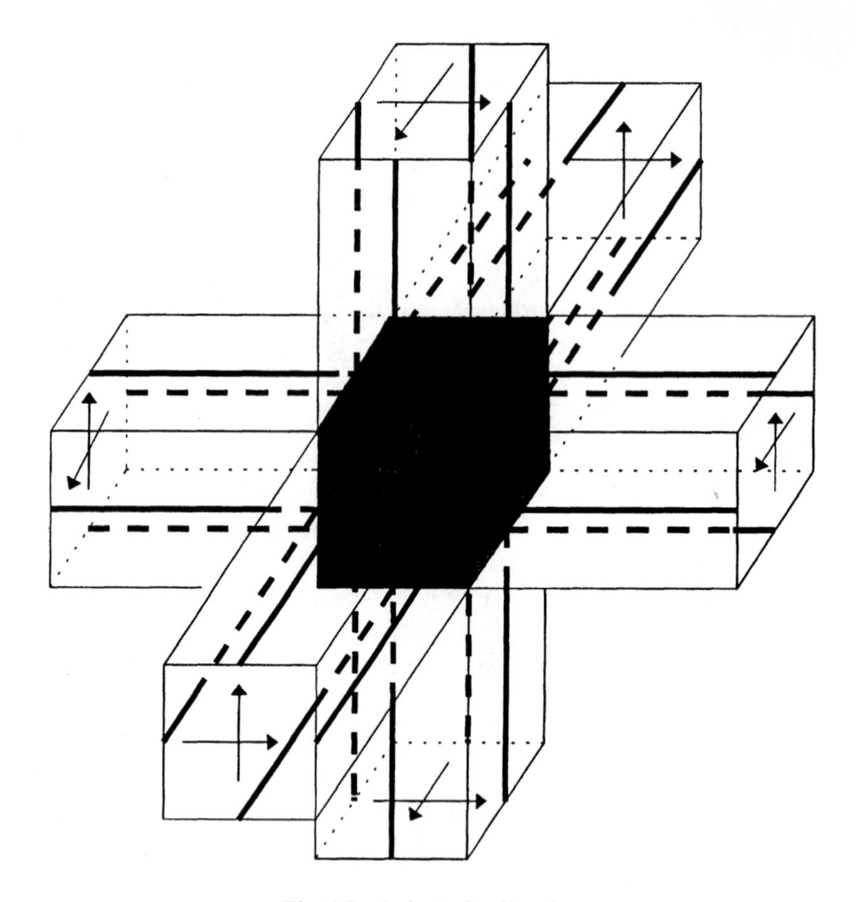

Fig. 6.7 A short-circuit node

time-step k + 1, is equal to the voltage reflected into port 12 of node (x, y–1, z) at time-step k; i.e.,

$$ {}_{k+1}V^{i}_{1}(x, y, z) = {}_{k}V^{r}_{12}(x, y-1, z) $$

Similar expressions apply for all other ports. An exception must be made for nodes that are adjacent to conducting, matched, or open-circuit boundaries. Such a node, next to a conducting boundary, is shown in Fig. 6.8. The new incident voltages on ports 10 and 11 are equal to minus the corresponding reflected voltages; i.e.,

$$ {}_{k+1}V^{i}_{10} = -{}_{k}V^{r}_{10} $$

$$ {}_{k+1}V^{i}_{11} = -{}_{k}V^{r}_{11} $$

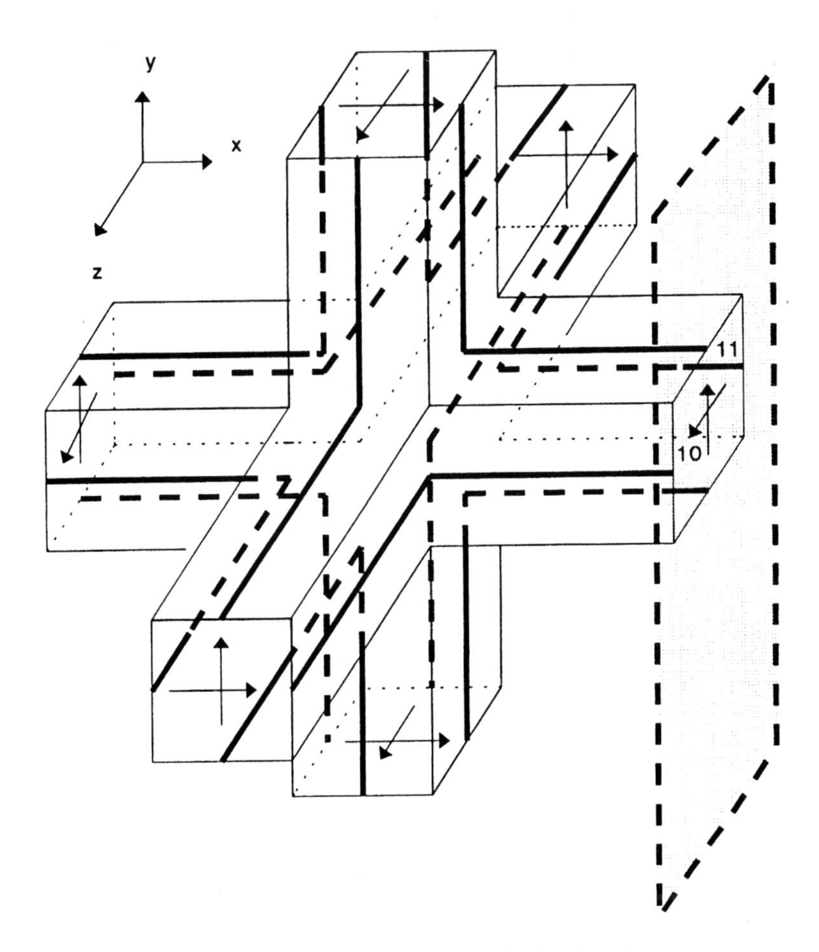

Fig. 6.8 An SCN adjacent to a conducting boundary

If this boundary has any other impedance value Z_b (corresponding to an open-circuit or matched condition), then the following procedure is applied:

$$_{k+1}V_n^i = \frac{Z_b - Z}{Z_b + Z} \, _kV_n^r \tag{6.17}$$

With the incident voltages at time-step $k + 1$ thus obtained, the scattered voltages may again be obtained at $k + 1$ from Equation (6.16), followed by further "connection" and "scattering" steps for as long as desired. The total number of time-steps required depends on the nature of the problem and the frequency resolution required. Typically, at least 1,000 steps are required to allow for several wave transits along the largest dimension of the problem.

At any stage during the simulation, the electric and magnetic field and other quantities are available and may form the output from the simulation. This is explained in Section 6.2.3. Results are given directly in the time domain. If required, they may be transformed to frequency-domain data using Fourier Transforms.

6.2.3 Output from an SCN Mesh

As already mentioned, all electromagnetic quantities may be obtained at any point in the mesh from the value of the incident voltage pulses.

Calculation of electric and magnetic fields. Let us first determine how the electric field component E_x may be obtained. From Equation (6.9),

$$E_x = -\frac{V_x}{\Delta \ell}$$

The total x-directed voltage is the average of the total voltage on ports 1, 2, 9, and 12; i.e.,

$$V_x = \frac{1}{4}\left[\left(V_1^i + V_1^r\right) + \left(V_2^i + V_2^r\right) + \left(V_9^i + V_9^r\right) + \left(V_{12}^i + V_{12}^r\right)\right]$$

It has already been shown, and it can also be confirmed directly using the scattering matrix, that for charge conservation, the sum of incident voltages to these ports is equal to the sum of reflected voltages. Hence,

$$V_x = \frac{1}{2}\left(V_1^i + V_2^i + V_9^i + V_{12}^i\right)$$

and

$$E_x = -\frac{V_1^i + V_2^i + V_9^i + V_{12}^i}{2\,\Delta \ell} \tag{6.18}$$

Using a similar procedure, the remaining field components may be calculated.

$$E_y = -\frac{V_3^i + V_4^i + V_{11}^i + V_8^i}{2\,\Delta \ell} \tag{6.19}$$

$$E_z = -\frac{V_5^i + V_6^i + V_7^i + V_{10}^i}{2\,\Delta \ell} \tag{6.20}$$

The magnetic field component, H_x, may be obtained from

$$H_x = \frac{I_x}{\Delta x}$$

The current, I_x, in turn, is calculated from the circuit shown in Fig. 6.4b, where each line is replaced by its Thevenin equivalent as shown in Fig. 6.9, to give

$$H_x = \frac{V_4^i + V_7^i - V_5^i - V_8^i}{2Z\,\Delta\ell} \tag{6.21}$$

Using similar reasoning, the following expressions are obtained:

$$H_y = \frac{-V_2^i + V_6^i + V_9^i - V_{10}^i}{2Z\,\Delta\ell} \tag{6.22}$$

$$H_z = \frac{V_1^i - V_3^i + V_{11}^i - V_{12}^i}{2Z\,\Delta\ell} \tag{6.23}$$

All electromagnetic field components may also be calculated, if desired, using the reflected voltages. The previous equations show the calculation of EM fields at the node. It is also possible to obtain the field

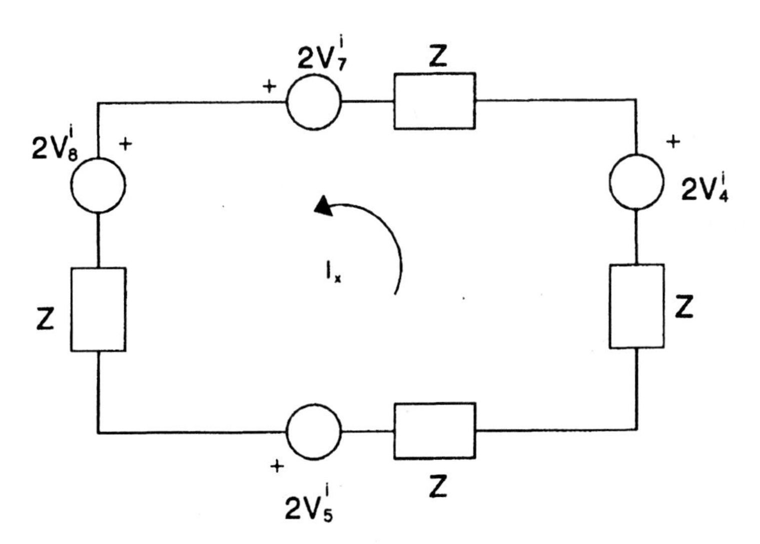

Fig. 6.9 Circuit for calculating the current contributing to the x-directed field

components more simply between nodes. As an illustration, the E_y and H_z components between nodes (x, y, z) and $(x + 1, y, z)$ are obtained from

$$E_y = -\frac{V_{11}^i(x, y, z) + V_3^i(x+1, y, z)}{\Delta y}$$

$$H_z = \frac{V_3^i(x+1, y, z) - V_{11}^i(x, y, z)}{\Delta z\, Zo}$$

An alternative method of deriving Equations (6.18) through (6.23) is described in Section 6.3.

Calculation of electric current. A quantity often required in simulations is the current flowing on conducting bodies. Currents may be obtained directly from incident voltages.

Let us consider the calculation of the current flowing along the y-direction on the perfectly conducting sheet shown in Fig. 6.8. The line driving current in this direction is number 11; hence, replacing it by its Thevenin equivalent gives the circuit shown in Fig. 6.10a. The current is

$$I_y = -\frac{2V^i_{11}}{Z}$$

If the conductor has resistance R, then the circuit shown in Fig. 6.10b applies, and

$$I_y = -\frac{2V^i_{11}}{R+Z}$$

The current flowing in the z-direction can be similarly obtained from port 10.

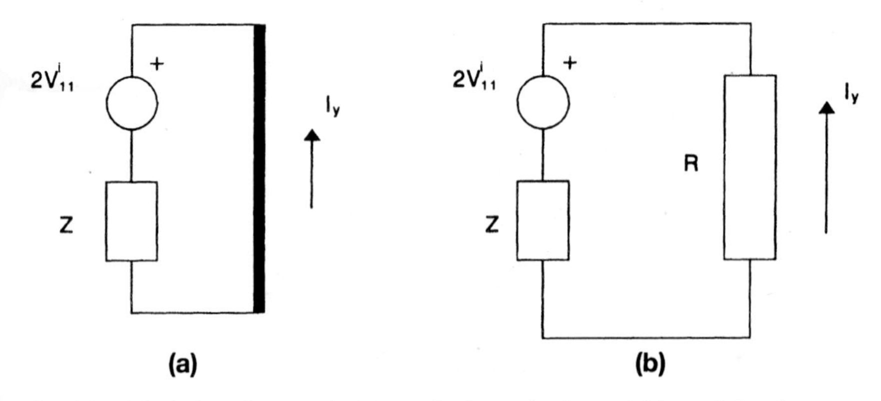

Fig. 6.10 Calculation of current in (a) a perfectly conducting and (b) a resistive sheet

The same technique may be applied to calculate the total current flowing in the z-direction in the short-circuit node shown in Fig. 6.7. This node may be part of a wire extending along z. The equivalent circuit is shown in Fig. 6.11. Using the parallel generator theorem and dividing by R gives the current

$$I_z = -2\frac{\left(V_5^i + V_6^i + V_7^i + V_{10}^i\right)}{4R + Z}$$

An alternative to this approach is to use Ampere's Law to calculate the current linked to a closed path surrounding the wire. Such a path is shown dotted in Fig. 6.12. Magnetic-field components, H, are calculated at each node using Equations (6.21) and (6.22). From Ampere's Law,

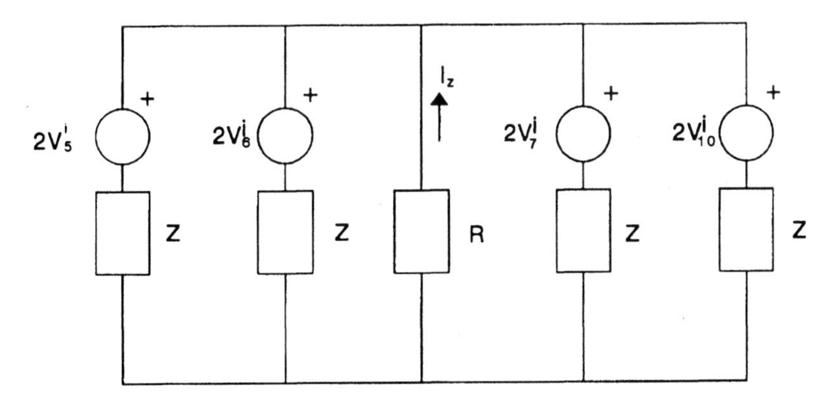

Fig. 6.11 Calculation of current in a short-circuit node

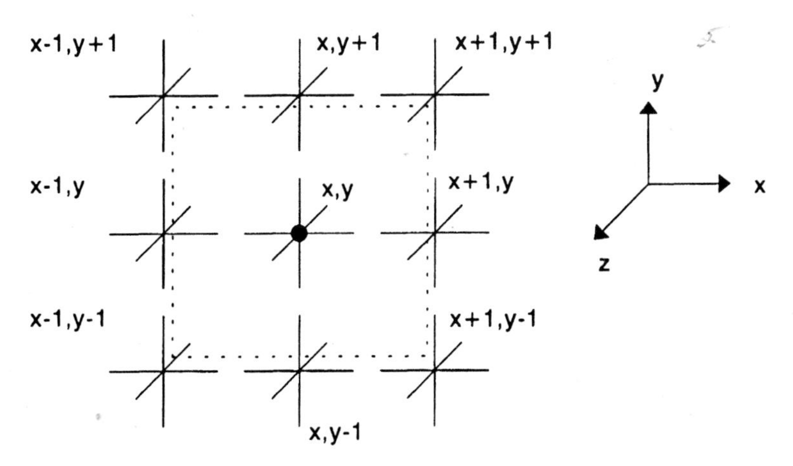

Fig. 6.12 Calculation of current in a short-circuit node at (x, y) using Ampere's Law

$$\int_c \mathbf{H}\, d\ell = I \tag{6.24}$$

where I is the current linked with curve c. Hence,

$$I_z = \frac{\Delta\ell}{2}H_x(x-1,y-1) + \Delta\ell H_x(x,y-1) + \frac{\Delta\ell}{2}H_x(x+1,y-1)$$

$$+\frac{\Delta\ell}{2}H_y(x+1,y-1) + \Delta\ell H_y(x+1,y) + \frac{\Delta\ell}{2}H_y(x+1,y+1)$$

$$-\frac{\Delta\ell}{2}H_x(x+1,y+1) - \Delta\ell H_x(x,y+1) - \frac{\Delta\ell}{2}H_x(x-1,y+1)$$

$$-\frac{\Delta\ell}{2}H_y(x-1,y+1) - \Delta\ell H_y(x-1,y) - \frac{\Delta\ell}{2}H_y(x-1,y-1)$$

Any closed path may be used for this calculation but it should be borne in mind that the current calculated includes, in addition to I_Z, any displacement current components.

Calculation of electric charge. The charge on a conductor may be calculated by applying Gauss's Law.

$$\int_s \mathbf{D}ds = Q \tag{6.25}$$

where Q is the charge enclosed by the surface s. The charge density on the surface shown in Fig. 6.8 is

$$q = -E_x \cdot \varepsilon_0 (\Delta\ell)^2$$

where E_x is calculated from (6.18). Using similar expressions, the charge on any conductor may be calculated.

6.2.4 Excitation in a SCN Mesh

In order to excite a particular field component at a node, it is necessary to inject voltages into the appropriate ports. In the case of E_X, pulses must be injected into ports 1, 2, 9, and 12, as suggested by Equation (6.18). Choosing E_0 as the desired field value and

$$V_1^i = V_2^i = V_9^i = V_{12}^i = -E_o\Delta\ell/2$$

and substituting into Equation (6.18) gives $E_x = E_0$ as desired. It should be noted that no other electric or magnetic field components are excited by this set of incident voltages. A similar approach may be adopted to excite other electric field components.

It is sometimes necessary in simulations to transfer charge between two conductors. This can be done simply by injecting pulses on ports and nodes forming a continuous path between the conductors. Care must be taken when exciting electric field components in free-space to minimize or take into account uncompensated charges that may be placed in the mesh as a result of the excitation process.

Magnetic field components may be similarly produced. To excite H_x it is necessary to inject voltages on ports 4, 5, 7, and 8, as suggested by Equation (6.21). Choosing H_0 as the desired value and

$$V_4^i = V_7^i = H_0 \, Z \, \Delta\ell/2$$

$$V_5^i = V_8^i = -H_0 \, Z \, \Delta\ell/2$$

and substituting into Equation (6.21) gives $H_x = H_0$ as desired. It can be confirmed by direct substitution into Equations (6.18) through (6.23) that no other field component is excited by this set of incident voltages.

It is sometimes necessary in simulations to inject a current into conductors. This can be done simply by injecting the necessary pulses to produce the desired value of

$$\int_c \mathbf{H} \, d\ell$$

on a path c surrounding the conductor [Ampere's Law, Equation (6.24)]. Other methods for injecting current are described in Section 9.4.

6.3 THE VARIABLE MESH SCN

There are two practical requirements that make it necessary to improve and extend the capabilities of the regular SCN described in Section 6.2.

First, it should be possible to use a fine mesh ($\Delta\ell$ small) only in areas of rapid field variation and thus save on computer requirements. This calls for a variable $\Delta\ell$ or what sometimes is referred to as a *graded mesh*.

Second, it should be possible to model nonuniform material properties such as different values of ε and μ. Modifications also must be made to model electric and magnetic losses.

Both these requirements must be met while maintaining synchronism (same time-step Δt throughout the mesh) and connectivity (one-to-one correspondence between transmission lines on adjacent ports), as discussed in connection with the 2D mesh (Section 5.1.2). The manner in which space may be divided to meet this requirement is shown in two dimensions in Fig. 6.13. It is therefore clear that the block of space modeled by each node is not necessarily a cube, and that the material properties in each block may be different. To accommodate these requirements, it is necessary to add stubs to the basic node shown in Fig. 6.3.

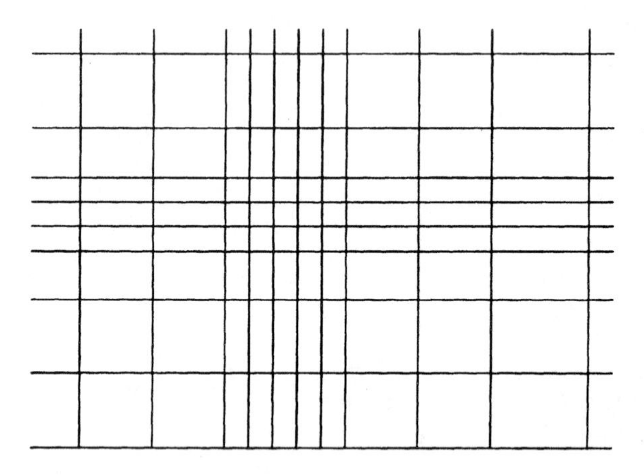

Fig. 6.13 A variable or graded mesh

6.3.1 The SCN with Capacitive and Inductive Stubs

To allow for variations in ε, μ, node shape, and dimensions, all 12 link transmission lines in the SCN node are chosen to model the same capacitance and inductance (C and L, respectively). These are chosen to represent the parameters of the background medium which, in most cases, is air (ε_0, μ_0). The impedance Z_0 of these lines is chosen the same throughout the mesh, and it has a value which in most cases corresponds to free space ($Z_0 = 377\ \Omega$). If the time step of the calculation is then chosen to be Δt, the propagation time on each line is $\Delta t/2$ and, hence, each line segment represents, according to Equation (5.29)

$$C = \frac{\Delta t/2}{Z_0},\ L = (\Delta t/2)\, Z_0$$

Clearly, depending on local material properties and the shape of the block modeled by the node, these values are to a greater or lesser extent underestimates of those required. These deficits in capacitance and inductance can be calculated and are then added to the SCN as stubs.

The required value for x-directed capacitance is

$$C_x = \varepsilon\frac{\Delta y\Delta z}{\Delta x}$$

The capacitance in this direction modeled so far is due to lines 1, 2, 9, and 12 and is therefore equal to $4C$. Hence, the deficiency is

$$C_x^s = \varepsilon\frac{\Delta y\Delta z}{\Delta x} - 4C \tag{6.26}$$

This capacitance is added to the node in the form of a stub of characteristic admittance

$$Y_x = \frac{2C_x^s}{\Delta t} = \frac{2\varepsilon\Delta y\Delta z}{\Delta t\Delta x} - \frac{8C}{\Delta t}$$

where the round trip on the stub is chosen to be equal to Δt to maintain synchronism. Substituting $C = \Delta t/2Z_0$ gives

$$Y_x = 2\varepsilon\frac{\Delta y\Delta z}{\Delta x\Delta t} - \frac{4}{Z_o} \tag{6.27}$$

Similar expressions may be obtained for the stubs that must be added to account for deficits in the y- and z-directions. It is customary to normalize the stub admittance to that of the background link lines, $Y_0 = 1/Z_0$; i.e.,

$$\hat{Y}_x = 2\varepsilon\frac{\Delta y\Delta z}{\Delta t\Delta x Y_0} - 4$$

Assuming that the background values are those for air ($Y_0 = \sqrt{\varepsilon_0/\mu_0}$) and substituting in the equation above gives the following expressions for $\hat{Y}_x$ and, by analogy, for the other capacitance stubs:

$$\hat{Y}_x = \frac{2\varepsilon_r}{u_o\Delta t}\,\frac{\Delta y\Delta z}{\Delta x} - 4$$

$$\hat{Y}_y = \frac{2\varepsilon_r}{u_o\Delta t}\,\frac{\Delta x\Delta z}{\Delta y} - 4$$

$$\hat{Y}_z = \frac{2\varepsilon_r}{u_o\Delta t}\,\frac{\Delta x\Delta y}{\Delta z} - 4 \tag{6.28}$$

where $u_o = 1/\sqrt{\mu_0\varepsilon_0}$.

Pulses scattered into the capacitive stubs are reflected after time $\Delta t/2$ from their open-circuit terminations and become incident onto the node at the next time-step.

A similar approach is adopted to calculate the inductance deficit. Let us consider first the deficit in inductance associated with current on the y-z plane.

The required value is

$$L_x = \mu\frac{\Delta y\Delta z}{\Delta x}$$

The modeled inductance is associated with lines 4, 5, 7, and 8 and is therefore equal to $4L$. Hence, the deficit is

$$L_x^s = \mu\frac{\Delta y\Delta z}{\Delta x} - 4L \tag{6.29}$$

This inductance is added to the node in the form of a stub of characteristic impedance

$$Z_x = \frac{2L_x^s}{\Delta t} = 2\mu\frac{\Delta y\Delta z}{\Delta t\Delta x} - \frac{8L}{\Delta t} = 2\mu\frac{\Delta y\Delta z}{\Delta t\Delta x} - 4Z_0$$

Normalizing as before to the background value Z_0 the following expressions are obtained for $\hat{Z}_x$ and, by analogy, for the other inductive stubs:

$$\hat{Z}_x = \frac{2\mu_r}{u_o\Delta t}\frac{\Delta y\Delta z}{\Delta x} - 4$$

$$\hat{Z}_y = \frac{2\mu_r}{u_o\Delta t}\frac{\Delta x\Delta z}{\Delta y} - 4$$

$$\hat{Z}_z = \frac{2\mu_r}{u_o\Delta t}\frac{\Delta x\Delta y}{\Delta z} - 4 \tag{6.30}$$

Pulses scattered into the inductive stubs are reflected after time $\Delta t/2$ from their short-circuit terminations and become incident onto the node at the next time-step.

The following two examples are given to show how parameters are chosen to model different media using nodes of various shapes.

EXAMPLE 6.1 Determine the stubs required to model a medium where $\varepsilon_r = \mu_r = 1$, using nodes representing a block of space with dimensions $\Delta x = \Delta y = 0.1$ m, $\Delta z = 0.3$m.

Solution: The characteristic impedance of free-space ($Z_0 = 377\Omega$) is chosen as the background. Then, from Equation (6.28),

$$\hat{Y}_x = \frac{2}{u_o \Delta t} \frac{0.1 \times 0.3}{0.1} - 4 = \frac{0.6}{u_o \Delta t} - 4$$

$$\hat{Y}_y = \frac{0.6}{u_o \Delta t} - 4$$

$$\hat{Y}_z = \frac{0.0666}{u_o \Delta t} - 4$$

and from (6.30),

$$\hat{Z}_x = \frac{2}{u_o \Delta t} \frac{0.1 \times 0.3}{0.1} - 4 = \frac{0.6}{u_o \Delta t} - 4$$

$$\hat{Z}_y = \frac{0.6}{u_o \Delta t} - 4$$

$$\hat{Z}_z = \frac{0.0666}{u_o \Delta t} - 4$$

For stability, it is necessary to ensure that all stubs represent real positive component values. Hence, the time-step must be chosen to ensure that all stub impedances and admittances are positive. From the previous expressions this means that

$$\frac{0.0666}{u_0 \Delta t} \geq 4, \quad \text{or}$$

$$\Delta t \leq 5.55 \times 10^{-11} s$$

■

EXAMPLE 6.2 Determine the stubs required to model a medium where $\varepsilon_r = 3$, $\mu_r = 1$, using nodes representing cubes $\Delta x = \Delta y = \Delta z = \Delta \ell$.

Solution: As before, using Z_0 as the background value, the following values are obtained from Equations (6.28) and (6.30):

$$\hat{Y}_x = \hat{Y}_y = \hat{Y}_z = \frac{6\Delta\ell}{u_o\Delta t} - 4$$

$$\hat{Z}_x = \hat{Z}_y = \hat{Z}_z = \frac{2\Delta\ell}{u_o\Delta t} - 4$$

For stability

$$\frac{2\Delta\ell}{u_o\Delta t} \geq 4, \quad \text{or}$$

$$\Delta t \leq \frac{\Delta\ell}{2u_o}$$

If the limiting value is chosen, the inductive stubs have zero impedance, and all capacitive stubs have a normalized admittance equal to eight.

These two examples show how parameters are chosen and the constraints on the time-step that are imposed to maintain stability. It should be pointed out that, in a large mesh, stability must be checked at each node and a suitable value of Δt chosen to prevent negative stub values anywhere in the mesh.

It remains now to examine how scattering takes place in this node.

■

Scattering in a stub-loaded SCN. The three capacitive and three inductive stubs are connected internally to the node and do not interact directly with adjacent nodes. Connection to neighboring nodes is, as for the regular SCN, through the 12 link lines. The scattering process, however, is affected by the presence of the stubs. The capacitive stubs are given port numbers 13 to 15, and the inductive stubs have port numbers 16 to 18. The scattering matrix now has dimensions 18 × 18. The non-zero elements of the scattering matrix may be identified using the same physical reasoning as outlined in Section 6.2. A pulse incident on port 1 may be partially reflected (amount a) and will couple to port 2 (amount b), port 9 (amount b) port 3 (amount d), port 11 (amount $-d$), and port 12 (amount c). In addition, since port 1 contributes to E_x and H_z, it will also couple with the x-directed capacitive stub (amount e) and z-directed inductive stub (amount f). In a similar manner, the other non-zero elements of S may be identified to produce the matrix shown in Equation (6.31). It must now be allowed that elements such as (1, 1), which is indicated as (a), may be different numerically from element (2, 2), also indicated as (a), since symmetry cannot be guaranteed in an irregularly shaped node.

S =	1	2	3	4	5	6	7	8	9	10	11	12	13	14	15	16	17	18
1	*a*	*b*	*d*						*b*		-*d*	*c*	*g*					*i*
2	*b*	*a*				*d*			*c*	-*d*		*b*	*g*				-*i*	
3	*d*		*a*	*b*				*b*			*c*	-*d*		*g*				-*i*
4			*b*	*a*	*d*		-*d*	*c*			*b*			*g*		*i*		
5				*d*	*a*	*b*	*c*	-*d*		*b*					*g*	-*i*		
6		*d*			*b*	*a*	*b*		-*d*	*c*					*g*		*i*	
7				-*d*	*c*	*b*	*a*	*d*		*b*					*g*	*i*		
8			*b*	*c*	-*d*		*d*	*a*			*b*			*g*		-*i*		
9	*b*	*c*				-*d*			*a*	*d*		*b*	*g*				*i*	
10		-*d*			*b*	*c*	*b*		*d*	*a*					*g*		-*i*	
11	-*d*		*c*	*b*				*b*			*a*	*d*		*g*				*i*
12	*c*	*b*	-*d*						*b*		*d*	*a*	*g*					-*i*
13	*e*	*e*							*e*			*e*	*h*					
14			*e*	*e*				*e*			*e*			*h*				
15					*e*	*e*	*e*			*e*					*h*			
16				*f*	-*f*		*f*	-*f*								*j*		
17		-*f*				*f*			*f*	-*f*							*j*	
18	*f*		-*f*								*f*	-*f*						*j*

(6.31)

The values of these elements may be calculated as before from Kirchhoff's Laws and from energy conservation. The unitary condition means, in this case, that the scattering matrix must obey the equation

$$\mathbf{S}^T \mathbf{Y} \mathbf{S} = \mathbf{Y}$$

where $\mathbf{Y}$ is an 18×18 diagonal matrix with elements the normalized admittance of the 12 link lines, 3 capacitive stubs, and 3 inductive stubs. It turns out, after some algebra, that the coefficients of the scattering matrix are [9]

$$a = \frac{-\hat{Y}}{2(4+\hat{Y})} + \frac{\hat{Z}}{2(4+\hat{Z})}$$

$$b = \frac{4}{2(4+\hat{Y})}$$

$$c = \frac{-\hat{Y}}{2(4+\hat{Y})} - \frac{\hat{Z}}{2(4+\hat{Z})}$$

$$d = \frac{4}{2(4+\hat{Z})}$$

$$e = b$$

$$f = \hat{Z}d$$

$$g = \hat{Y}b$$

$$h = \frac{\hat{Y}-4}{\hat{Y}+4}$$

$$i = d$$

$$j = \frac{4-\hat{Z}}{4+\hat{Z}} \tag{6.32}$$

The values of $\hat{Y}$ and $\hat{Z}$ used in these formulae are chosen to correspond to the relevant stubs. For example, element $d_{6,9}$ represents coupling between ports 6 and 9. Port 6 is associated with E_z and H_y, while port 9 with E_x and H_y. Hence, in the formulae for d, the value for $\hat{Z}$ must be that corresponding to H_y; i.e.,

$$d_{6,9} = \frac{4}{2(4+\hat{Z}_y)}$$

Similarly, ports 2 and 9 are associated with E_x and H_y; hence, in calculating $C_{2,9}$, $\hat{Y}_x$ and $\hat{Z}_y$ must be used.

Calculation of EM fields in a stubbed SCN. Let us now examine how E_x may be calculated. The procedure adopted in Section 6.2.3 must now be modified to take account of the influence of the x-directed capacitive stub (port 13). The capacitance of each link line is $Y_0 \Delta t/2$ (see Section 6.3.1), and the capacitance of the stub 13 is $Y_x \Delta t/2 = \hat{Y}_x Y_0 \Delta t/2$. Hence, the total incident charge on all the ports associated with E_x is

$$\frac{Y_0 \Delta t}{2}\left(V_1^i + V_2^i + V_9^i + V_{12}^i\right) + Y_0 \frac{\Delta t}{2} \hat{Y}_x V_{13}^i$$

From charge conservation, the total charge leaving these lines should be the same as above; hence, in total, the charge is equal to twice the above value. The total capacitance of these lines is

$$4Y_0 \frac{\Delta t}{2} + \hat{Y}_x Y_0 \frac{\Delta t}{2}$$

Dividing charge by capacitance gives the x-directed voltage

$$V_x = \frac{2\left(V_1^i + V_2^i + V_9^i + V_{12}^i + \hat{Y}_x V_{13}^i \right)}{4 + \hat{Y}_x}$$

The electric field is then equal to $-V_X/\Delta x$; hence,

$$E_x = -\frac{2\left(V_1^i + V_2^i + V_9^i + V_{12}^i + \hat{Y}_x V_{13}^i \right)}{\Delta x\,(4 + \hat{Y}_x)} \qquad (6.33)$$

Similar expressions apply for the other two electric field components.

Component H_x may be obtained by considering contributions from ports responsible for currents flowing in the y-z plane. These are shown in Fig. 6.14, where the inductive stub associated with H_x has been added with positive reference voltage opposing the positive reference current I_x.

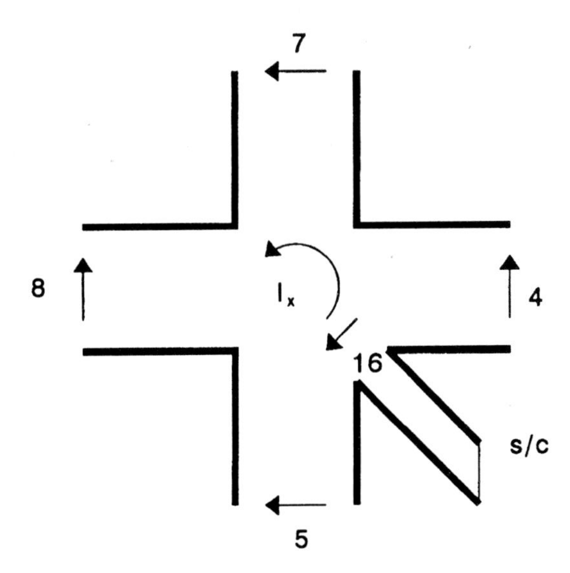

Fig. 6.14 Circuit for calculating the x-component of the magnetic field in a node with stubs

From this circuit, using Thevenin equivalents, the current is obtained, and since $H_x = I_x/\Delta X$,

$$H_x = \frac{2\left(V_4^i - V_5^i + V_7^i - V_8^i - V_{16}^i\right)}{\Delta x\left(4Z_0 + Z_0\hat{Z}_z\right)} \tag{6.34}$$

Similar expressions are obtained for the other magnetic field components.

Excitation in a stubbed SCN. Excitation is similar to that for the unstubbed node. For example, if E_x only is to be excited, the following pulses need to be injected:

$$V_1^i = V_2^i = V_9^i = V_{12}^i = V_{13}^i = \frac{\Delta x}{2}E_0$$

where E_0 is the desired value of E_x in volts per meter. More details and formulae may be found in Ref. [9].

6.3.2 The SCN with Capacitive, Inductive, and Lossy Stubs

Losses may be incorporated by adding more stubs to the node described in the previous section. These "lossy" stubs may be regarded as infinitely long or, equivalently, as terminated by their own characteristic impedance. In either case, any energy scattered into these stubs is absorbed, and there are no pulses coming from them that are incident on the node. Three lossy stubs associated with electric losses in the x-, y-, and z-directions (ports 19–21) and another three associated with magnetic losses in the x-, y-, and z-directions (ports 22–24) may be introduced to produce the most versatile node capable of dealing with general inhomogeneous anisotropic media. It is, however, rare that such a general node is required. Such a general node requires substantial computation, and special algorithms have been developed to increase numerical efficiency. These algorithms should be used—especially with stubbed nodes [15].

The introduction of electric and magnetic losses is described in Refs. [16] and [17], respectively. Since no incident voltages are coming from the lossy stubs, the scattering matrix has 18 columns, 24 rows. Normally, the pulses scattered into the lossy nodes need not be explicitly calculated, and then **S** is an 18 × 18 matrix. Row 19 of **S** consists of elements representing ports coupled to losses associated with E_x. Hence, coupling is represented by elements 1, 2, 9, 12 (amount k), and 13 (amount p).

Similarly, row 20 has non-zero elements 3, 4, 8, 11 (amount k), and 14 (amount p).

For row 21, the non-zero elements are 5, 6, 7, 10 (amount k), and 15 (amount p).

Row 22 consists of elements representing losses associated with H_x. Non-zero elements are 4, 7 (amount k'), 5, 8 (amount $-k'$), and 16 (amount p'). Row 23 has non-zero elements 6, 8 (amount k'), 2, 10 (amount $-k'$), and 17 (amount p').

Finally, row 24 has elements 1, 11 (amount k'), 3, 12 (amount p'), and 18 (amount p$'$).

Stubs 19 through 21 have normalized conductances $\hat{G}_x$, $\hat{G}_y$, and $\hat{G}_z$, respectively. Similarly, stubs 22 to 24 have normalized resistances $\hat{R}_x$, $\hat{R}_y$, and $\hat{R}_z$, respectively. Following a procedure identical to that described in the previous section, the following values for the elements of the 18×18 scattering matrix are obtained:

$$a = -\frac{\hat{Y}+\hat{G}}{2(\hat{Y}+\hat{G}+4)} + \frac{\hat{Z}+\hat{R}}{2(\hat{Z}+\hat{R}+4)}$$

$$b = \frac{4}{2(\hat{Y}+\hat{G}+4)}$$

$$c = -\frac{\hat{Y}+\hat{G}}{2(\hat{Y}+\hat{G}+4)} - \frac{\hat{Z}+\hat{R}}{2(\hat{Z}+\hat{R}+4)}$$

$$d = \frac{4}{2(\hat{Z}+\hat{R}+4)}$$

$$e = b$$

$$f = \hat{Z}d$$

$$g = \hat{Y}d$$

$$h = \frac{\hat{Y}-\hat{G}-4}{\hat{Y}+\hat{G}+4}$$

$$i = d$$

$$j = \frac{4-\hat{R}-\hat{Z}}{4+\hat{R}+\hat{Z}} \tag{6.35}$$

As for the lossless node, the appropriate values of $\hat{Z}$, $\hat{Y}$, $\hat{G}$, and $\hat{R}$ must be used. As an illustration, $C_{2,9}$ represents coupling between ports 2 and 9, both of which are associated with E_x and H_y. Hence,

$$C_{2,9} = -\frac{\hat{Y}_x + \hat{G}_x}{2(\hat{Y}_x + \hat{G}_x + 4)} - \frac{\hat{Z}_y + \hat{R}_y}{2(\hat{Z}_y + \hat{R}_y + 4)}$$

The lossy stub parameters are calculated from the following formula:

$$G_x = \sigma_{ex}\frac{\Delta y \Delta z}{\Delta x}$$

hence the normalized value is

$$\hat{G}_x = \sigma_{ex}\frac{\Delta y \Delta z}{\Delta x Y_0} \tag{6.36}$$

Parameter σ_{ex} is the conductivity associated with electric losses in the x-direction. The complex dielectric constant, conductivity, and loss tangent are related by the expressions

$$\varepsilon^* = \varepsilon_r\varepsilon_0 - j\varepsilon'' = \varepsilon_r\varepsilon_0 - j\frac{\sigma_e}{\omega}$$

$$\tan\delta_e = \frac{\sigma_e}{\omega\varepsilon_r\varepsilon_0} \tag{6.37}$$

Similar expressions apply for G_y and G_z.

To account for magnetic losses associated with H_x, a resistance R_x must be added in series with L_x. The power loss is then $P_{mx} = I_x^2 R_x$. From the phasor diagram for this circuit, shown in Fig. 6.15, a loss-tangent may be associated with magnetic losses

$$\tan\delta_{mx} = \frac{V_R}{V_L} = \frac{R_x}{\omega L_x}$$

Therefore,

$$R_x = \omega L_x \tan\delta_{mx} = (\omega\mu\tan\delta_{mx})\frac{\Delta y \Delta z}{\Delta x}$$

The complex magnetic permeability of the medium is

$$\mu^* = \mu_r\mu_0 - j\mu'' = \mu_r\mu_0 - j\frac{\sigma_m}{\omega} \tag{6.38}$$

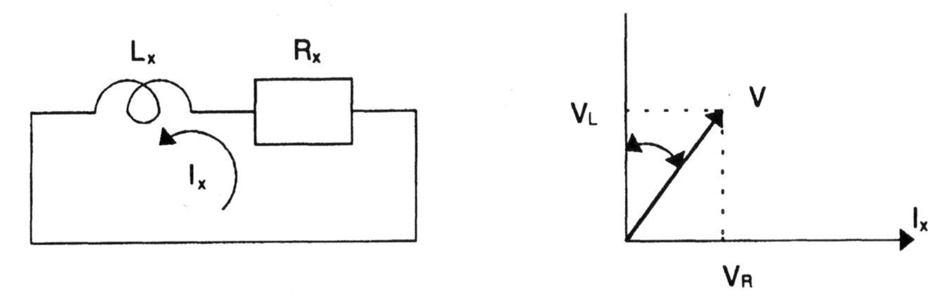

Fig. 6.15 Circuit for the characterization of magnetic losses

Hence,

$$\tan \delta_m = \frac{\sigma_m}{\omega \mu_r \mu_0}$$

and substituting in the expression for R_x gives

$$R_x = \sigma_{mx} \frac{\Delta y \Delta z}{\Delta x}$$

The normalized value is

$$\hat{R}_x = \sigma_{mx} \frac{\Delta y \Delta z}{\Delta x Z_0} \tag{6.39}$$

Similar expressions are obtained for R_y and R_z.

An alternative formulation which gives a different frequency dependence of the real and imaginary parts of the complex magnetic permeability is described in Ref. [18]. Magnetic losses are accounted for by introducing a resistance in parallel with the stub inductance. The resulting variation in μ^* with frequency is closer to that observed in ferrite materials. Details of the modified scattering matrix may be found in the reference mentioned above.

6.3.3 Non-Cartesian SCN

Modeling so far has been limited to the development of nodes to model cubic shaped blocks of space—discretization in space took place in a cartesian grid. There are problems, however, such as those with cylindrical or spherical symmetry, where it would be profitable to use nodes that model directly non-cubic shaped blocks. The development of variable or graded mesh [4, 5] has made possible modeling in a non-cartesian grid. Modeling in a cylindrical mesh has been described in Ref. [19], and TLM in a general orthogonal curvilinear mesh in two dimensions was described in Ref. [20].

The advantages of using grids other than cartesian is that, in certain cases, it may be possible to describe problem boundaries more accurately. A curved boundary would have to be described in a step-wise fashion if a rectangular grid is used. Another advantage of a curvilinear mesh is that a saving in computer storage may be achieved by exploiting problem symmetry. As an illustration, in a case of cylindrical symmetry, it is only necessary to model a portion of the problem shaped like a slice of cake (see Fig. 6.16a). Planes ABCD and ABFE are planes of symmetry and represent an open-circuit boundary condition for the layer of nodes representing the volume of the slice.

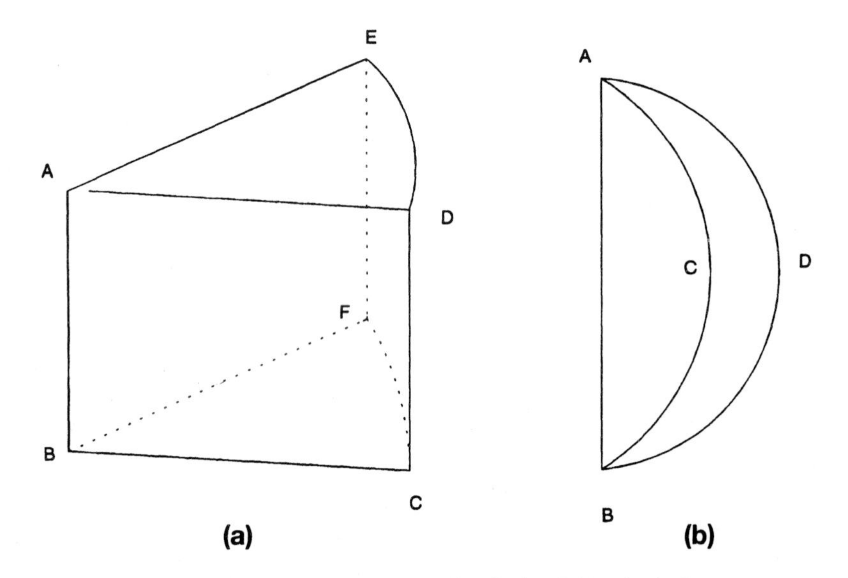

Fig. 6.16 A slice in problems with (a) cylindrical and (b) spherical symmetry

Similarly, in a problem with spherical symmetry, it is only necessary to model a portion, shaped like a slice of orange, as shown in Fig. 6.16b. Surfaces ABC and ABD form an open-circuit boundary condition for the layer of nodes modeling the slice.

The modifications necessary to adapt the SCN for use in a cylindrical element are described in the following text.

The coordinate system used and the port designations are shown in Fig. 6.17. Simulation proceeds exactly as for a SCN with stubs in a cartesian grid. The only modification involves the calculation of stub parameters where account must be taken of the details of the new geometry.

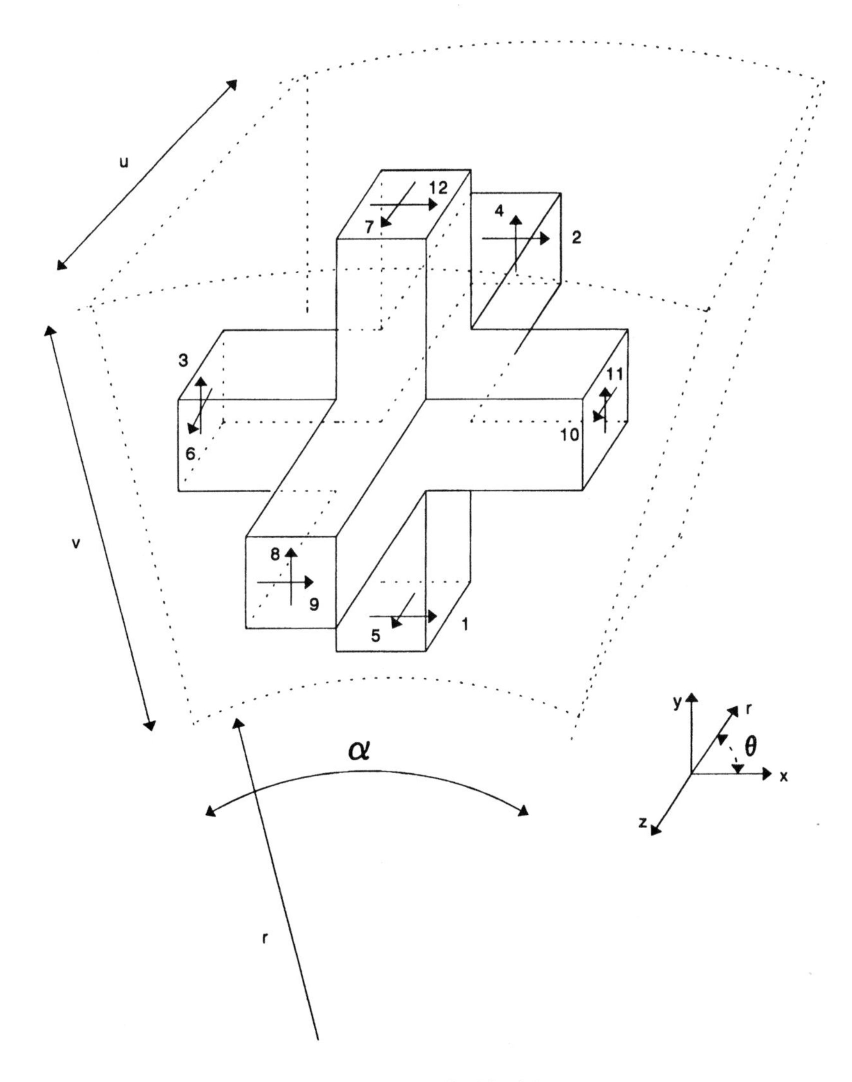

Fig. 6.17 A cylindrical SCN

The capacitance that must be modeled is

$$C_r = \varepsilon \frac{(r\alpha)\, u}{v}$$

$$C_z = \varepsilon \frac{(r\alpha)\, v}{u}$$

$$C_\theta = \varepsilon \frac{uv}{(r\alpha)} \tag{6.40}$$

If each link line has a characteristic impedance Z_0, then the modeled capacitance in each direction is

$$4\frac{\Delta t/2}{Z_0} = 2\Delta t/Z_0$$

Hence, the deficit in the r-direction is

$$C_r^s = C_r - 2\Delta t/Z_0$$

The admittance of the r-directed stub is

$$Y_r = \frac{2C_r^s}{\Delta t}$$

and the normalized value after substitution is

$$\hat{Y}_r = \frac{2C_r^s/(\Delta t)}{Y_0} = 2\varepsilon_r \frac{ur\alpha}{u_0 \Delta tv} - 4 \tag{6.41}$$

Similar expressions apply for the remaining two capacitive stubs.

The desired inductance associated with H_r is

$$L_r = \mu\frac{(r\alpha)u}{v}$$

The inductance modeled in this direction by the four link lines is $4\,Z_0\,\Delta t/2$. Hence, the deficit is

$$L_r^s = \mu\frac{(r\alpha)u}{v} - 2Z_0\Delta t$$

The normalized inductive stub impedance is then

$$\hat{Z}_r = 2\mu_r \frac{(r\alpha)u}{u_o\,\Delta t\,v} - 4 \tag{6.42}$$

Similar expressions apply for the remaining two inductive stubs. As mentioned before in connection with the variable cartesian mesh, the choice of time-step must be such that all stubs everywhere in the mesh represent positive circuit components. This requirement can force the time-step to very low values and hence lengthen the computation significantly. However, the use of the cylindrical mesh can bring worthwhile benefits in certain types of problem [21].

6.4 THE HYBRID SCN

A variable mesh using the stubbed SCN was described in Section 6.3.1. Examination of the constraints on the maximum permissible time-step imposed by the requirements of stability (positive stub impedances) shows that it is determined by the ratio between the largest and smallest grid dimensions. The implications of this are that, in a large problem where large ratios are used to minimize storage, the resulting time-step is very small, thus requiring long computational runs. Benefits resulting from reductions in storage are diminished by the corresponding increases in run time. These difficulties can be overcome if the condition inherent in the SCN development—that all link lines have the same characteristic impedance—is relaxed. This idea was explored in Ref. [22] and implemented for the expanded node in Ref. [23]. Its implementation for the SCN was described in Refs. [24–26]. The resulting node, referred to as the *hybrid symmetrical condensed node* (HSCN) consists of link lines that model all the required inductance at the node. The capacitance modeled by these lines is chosen so that synchronism is maintained. Any deficit in capacitance is accounted for by introducing three capacitive stubs. Hence, compared to the stubbed SCN, three fewer stubs are required. This means reduced storage plus better dispersion properties. More importantly, the HSCN can be operated at a higher time-step than the stubbed SCN. Parameter determination and scattering in the HSCN are described in the following text.

The 3D node consists of three networks, as was shown schematically in Fig. 6.4 and repeated in more detail in Fig. 6.18. Currents flowing in the network shown in Fig 6.18a are related to H_z; hence, the admittance of lines 1, 3, 11, 12 is chosen to be the same and is designated by Y_z. The total inductance required is

$$L_z = \mu\frac{\Delta x \Delta y}{\Delta z} = L_{dxy}\,\Delta y + L_{dyx}\,\Delta x \tag{6.43}$$

where L_{dxy} indicates the inductance per unit length (first subscript d) for a line polarized in the x-direction (second subscript) and for propagation in the y-direction (third subscript). Similarly, from Fig. 6.18b and c,

$$L_x = \mu\frac{\Delta y \Delta z}{\Delta x} = L_{dzy}\,\Delta y + L_{dyz}\,\Delta z$$

$$L_y = \mu\frac{\Delta x \Delta z}{\Delta y} = L_{dxz}\,\Delta z + L_{dzx}\,\Delta x \tag{6.44}$$

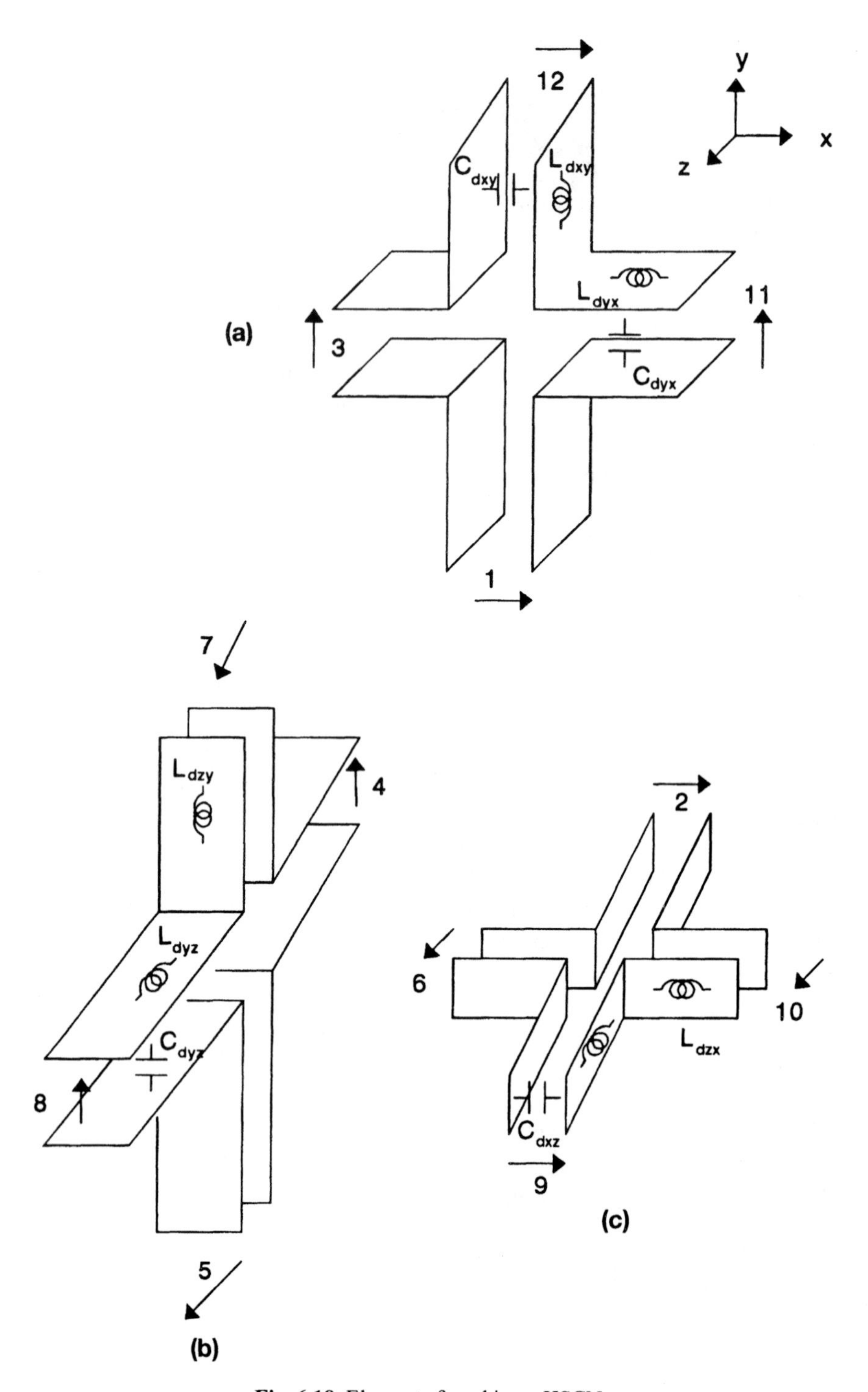

Fig. 6.18 Elements found in an HSCN

Let ΔL be the smallest of all nodal dimensions Δx, Δy, Δz. The time-step is chosen to be

$$\Delta t = \frac{\Delta L}{u_{TL}} = \frac{\Delta L}{h}\sqrt{\mu_0 \varepsilon_0} \tag{6.45}$$

where u_{TL} is the velocity of propagation on each line, and h is a constant to be determined.

On a line of length $\Delta \ell$, with propagation time Δt and inductance and capacitance per unit length L_d, and C_d, respectively, the velocity of propagation is

$$u_{TL} = \frac{\Delta \ell}{\Delta t} = \frac{1}{\sqrt{L_d C_d}}$$

Hence,

$$C_d = \frac{(\Delta t)^2}{(\Delta \ell)^2} \frac{1}{L_d}$$

Substituting for Δt from Equation (6.45) gives

$$C_d = \frac{(\Delta L)^2}{h^2} \mu_0 \varepsilon_0 \frac{1}{(\Delta L)^2 L_d} \tag{6.46}$$

Since we require all lines contributing to the inductance associated with a particular magnetic field component to have the same impedance (i.e.; that for H_z, $Z_1 = Z_3 = Z_{11} = Z_{12}$), this in turn requires that

$$\sqrt{\frac{L_{dxy}}{C_{dxy}}} = \sqrt{\frac{L_{dyx}}{C_{dyx}}}$$

or that

$$\frac{L_{dxy}\Delta y}{C_{dxy}\Delta y} = \frac{L_{dyx}\Delta x}{C_{dyx}\Delta x}$$

This condition can be met if

$$L_{dxy}\Delta y = L_{dyx}\Delta x$$

$$C_{dxy}\Delta y = C_{dyx}\Delta x \tag{6.47}$$

Combining the first of these equations with Equation (6.43) gives

$$L_{dxy} = \frac{\mu \Delta x}{2\Delta z}$$

$$L_{dyx} = \frac{\mu \Delta y}{2\Delta z} \tag{6.48}$$

Similar reasoning for the lines associated with the other two field components gives

$$L_{dxz} = \frac{\mu \Delta x}{2\Delta y}$$

$$L_{dzx} = \frac{\mu \Delta x}{2\Delta y} \tag{6.49}$$

$$L_{dyz} = \frac{\mu \Delta y}{2\Delta z}$$

$$L_{dzy} = \frac{\mu \Delta z}{2\Delta x} \tag{6.50}$$

Equations (6.48) through (6.50) determine the inductance per unit length for each link line in the HSCN. These values account for all the inductance required at the node.

Let us now turn attention to the capacitance. The capacitance deficit in the x-direction is

$$C_x^s = \varepsilon \frac{\Delta y \Delta z}{\Delta x} - (C_{dxy}\, \Delta y + C_{dxz}\, \Delta z) \tag{6.51}$$

where the capacitance per unit length must be chosen so that the time it takes to travel the length of the line Δy is Δt. It follows from Equation (6.46) that

$$C_{dxy} = \frac{(\Delta L)^2 \mu_0 \varepsilon_o}{h^2 L_{dxy} (\Delta y)^2} \tag{6.52}$$

Substituting for L_{dxy} from Equation (6.48) gives

$$C_{dxy} = \frac{2\,(\Delta L)^2 \mu_0\, \varepsilon_0\,(\Delta z)^2}{h^2\, \mu\, \Delta x\, \Delta y\, \Delta z}$$

A formula similar to this is obtained for C_{dxz}, and both these expressions are substituted in Equation (6.51) to give

$$C_x^s = \frac{\Delta y \Delta z}{\Delta x} - \frac{2\,(\Delta L)^2 \mu_0 \varepsilon_o}{h^2\, \mu\, \Delta x\, \Delta y\, \Delta z}\,[\,(\Delta z)^2 + (\Delta y)^2] \tag{6.53}$$

Stability requires that $C_x^s \geq 0$. Hence, from Equation (6.53), the constant h must be chosen so that

$$h^2 \geq \frac{2\,(\Delta L)^2\,[\,(\Delta y^2) + (\Delta z)^2]}{\mu_r \varepsilon_r\,(\Delta y^2)\,(\Delta z)^2} \tag{6.54}$$

The worst case corresponds to $\mu_r = \varepsilon_r = 1$, and substituting in Equation (6.54),

$$h^2 \geq 2\,(\Delta L)^2 \left[\frac{1}{(\Delta y)^2} + \frac{1}{(\Delta z)^2}\right]$$

The most conservative estimation of h can be obtained by replacing Δy and Δz in the expression above by the minimum nodal dimension ΔL, which gives

$$h \geq 2 \tag{6.55}$$

It follows that if h is chosen according to Equation (6.55), stability is maintained under all circumstances.

The next step is the calculation of the link line admittance starting with Y_x for lines 4, 5, 7, and 8.

$$Y_x = \sqrt{\frac{C_{dyz}}{L_{dyz}}} = \frac{\Delta x\, \Delta L\, \sqrt{\mu_0 \varepsilon_0}}{\mu\, \Delta y\, \Delta z}$$

where Equation (6.52) has been used to substitute for the capacitance. The normalized admittance is

$$\hat{Y}_x = \frac{Y_x}{Y_0} = \frac{\Delta x\, \Delta L}{\mu_r\, \Delta y\, \Delta z} \tag{6.56}$$

Similar expressions apply for the normalized admittance of lines 2, 6, 9 and 10

$$\hat{Y}_y = \frac{\Delta L\ \Delta y}{\mu_r\ \Delta x\ \Delta z} \tag{6.57}$$

and for lines 1, 3, 11, and 12.

$$\hat{Y}_z = \frac{\Delta L\ \Delta z}{\mu_r\ \Delta y\ \Delta x} \tag{6.58}$$

The normalized admittance of the x-directed capacitive stub is

$$\hat{Y}_x^s = \frac{Y_x^s}{Y_0} = \frac{2C_x^s}{\Delta t\ Y_0}$$

Substituting for the deficit in capacitance C_x^s from Equation (6.53) and choosing $h = 2$ gives

$$\hat{Y}_x^s = \frac{4\varepsilon_r\ \Delta y\ \Delta z}{\Delta L\ \Delta x} - \frac{2\Delta L\,[\,(\Delta y)^2 + (\Delta z)^2]}{\mu_r\ \Delta x\ \Delta y\ \Delta z} \tag{6.59}$$

Similar expressions are obtained for the remaining two stubs

$$\hat{Y}_y^s = \frac{4\varepsilon_r\ \Delta x\ \Delta z}{\Delta L\ \Delta y} - \frac{2\Delta L\,[\,(\Delta x)^2 + (\Delta z)^2]}{\mu_r\ \Delta x\ \Delta y\ \Delta z} \tag{6.60}$$

$$\hat{Y}_z^s = \frac{4\varepsilon_r\ \Delta x\ \Delta y}{\Delta L\ \Delta z} - \frac{2\Delta L\,[\,(\Delta x)^2 + (\Delta y)^2]}{\mu_r\ \Delta x\ \Delta y\ \Delta z} \tag{6.61}$$

Scattering in the HSCN. The scattering matrix for this node is given for the very general case of electric and magnetic losses and current density source terms. Ports 1 through 12 correspond to the link lines, ports 13 through 15 to the capacitive stubs, and ports 16 through 18 are the source connections to the node. The reflected voltages into the lossy stubs need not be calculated, as the energy associated with them is absorbed and no voltages incident to the node originate from them. The source supplies incident pulses only. The pulses reflected into the lossy nodes may be calculated, if desired, but since this is not required in most cases, the scattering matrix is given in its most efficient compact form. It thus has 15 rows and 18 columns. If source terms are not required, a 15 × 15 scattering matrix results. A pulse incident on port 1 is associated with E_x and H_z and, as explained before, couples with ports 1, 2, 3, 9, 11, 12, 13, the x-directed electric loss stub, and the z-directed magnetic loss stub. The part of the scattering matrix of direct interest is the first column of **S** in Equation (6.62).

		Y_ℓ	y	z	x	z	y	x	y	z	z	x	x	y						
		Y_t	z	y	z	x	x	y	x	x	y	y	z	z						
		Y_S	x	x	y	y	z	z	z	y	x	z	y	x	x	y	z	x	y	z
R_t	G_S		1	2	3	4	5	6	7	8	9	10	11	12	13	14	15	16	17	18
z	x	1	a	b	d						b		$-d$	c	g			k		
y	x	2	b	a				d			c	$-d$		b	g			k		
z	y	3	d		a	b				b			c	$-d$		g			k	
x	y	4			b	a	d		$-d$	c			b			g			k	
x	z	5				d	a	b	c	$-d$		b					g			k
y	z	6			d		b	a	b		$-d$	c					g			k
x	z	7				$-d$	c	b	a	d		b					g			k
x	y	8			b	c	$-d$		d	a			b			g			k	
y	x	9	b	c				$-d$			a	d		b	g			k		
y	z	10		$-d$			b	c	b		d	a					g			k
z	y	11	$-d$		c	b				b			a	d		g			k	
z	x	12	c	b	$-d$						b		d	a	g			k		
	x	13	b	b							b			b	h			k		
	y	14			b	b				b			b			h			k	
	z	15					b	b	b			b					h			k

(6.62)

Similar reasoning may be employed to identify the remaining non-zero elements of Equation (6.62). The form of the scattering coefficient is given below [24, 26].

$$a = \frac{\hat{Y}_S + \hat{G}_S + 2\,(\hat{Y}_\ell - \hat{Y}_t)}{2\,[\hat{Y}_S + \hat{G}_S + 2\,(\hat{Y}_\ell + \hat{Y}_t)\,]} + \frac{\hat{R}_t \hat{Y}_t}{2\,(\hat{R}_t \hat{Y}_t + 4)}$$

$$b = \frac{2\,\hat{Y}_t}{\hat{Y}_S + \hat{G}_S + 2\,(\hat{Y}_\ell + \hat{Y}_t)}$$

$$c = -\frac{\hat{Y}_s + \hat{G}_s + 2(\hat{Y}_\ell - \hat{Y}_t)}{2[\hat{Y}_s + \hat{G}_s + 2(\hat{Y}_\ell + \hat{Y}_t)]} - \frac{\hat{R}_t \hat{Y}_t}{2(\hat{R}_t \hat{Y}_t + 4)}$$

$$d = \frac{2}{\hat{R}_t \hat{Y}_t + 4}$$

$$g = b\frac{\hat{Y}_s}{\hat{Y}_t}$$

$$h = \frac{\hat{Y}_s - \hat{G}_s - 2(\hat{Y}_\ell - \hat{Y}_t)}{\hat{Y}_s + \hat{G}_s + 2(\hat{Y}_\ell + \hat{Y}_t)}$$

$$k = \frac{1}{\hat{Y}_s + \hat{G}_s + 2(\hat{Y}_\ell + \hat{Y}_t)} \tag{6.63}$$

where $\hat{Y}_\ell$, $\hat{Y}_t$ represents the admittances of the link lines normalized to the free-space admittance. Depending on the line selected, they have values obtained from Equations (6.56) through (6.58).

The open-circuit stub admittances are represented by $\hat{Y}_s$ and, depending on the particular stub selected, they have values obtained from Equations (6.59) through (6.61). Electric losses are represented by $\hat{G}_s$, with values obtained from expressions of the form shown in Equation (6.36). Similarly, magnetic losses are represented by $\hat{R}_t$, with values obtained from Equation (6.39) and its equivalents for the y- and z-directions.

As an illustration, element (1, 1) of the scattering matrix is calculated. Using the key shown in Equation (6.62), the following values must be used for the calculation of $d_{1,1}$:

$$\hat{Y}_\ell = \hat{Y}_y \qquad \hat{Y}_s = \hat{Y}_x^s \qquad \hat{G}_s = \hat{G}_x$$

$$\hat{Y}_t = \hat{Y}_z \qquad \qquad \hat{R}_t = \hat{R}_z$$

Similarly, for element (11, 3) of **S**, the following choices in the calculation of $c_{11,3}$ must be made:

$$\hat{Y}_\ell = \hat{Y}_x \qquad \hat{Y}_s = \hat{Y}_y^s \qquad \hat{G}_s = \hat{G}_y$$

$$\hat{Y}_t = \hat{Y}_z \qquad \qquad \hat{R}_t = \hat{R}_z$$

The voltage pulses incident on ports 16 through 18 represent the source current density as follows:

$$V_{16}^{i} = j_x Z_0 \Delta y \Delta z$$

$$V_{17}^{i} = j_y Z_0 \Delta x \Delta z$$

$$V_{18}^{i} = j_z Z_0 \Delta x \Delta y \tag{6.64}$$

The connection procedure for the HSCN is similar to that for the SCN, but now account must be taken of the different values of the characteristic impedance of the link lines.

Reflection at external boundaries in an HSCN mesh. Let us consider the termination of a y-polarized line such as line 11. The impedance associated with this block of space is

$$Z = \sqrt{\frac{L_z}{C_y}} = Z_0 \sqrt{\frac{\mu_r}{\varepsilon_r} \frac{\Delta y}{\Delta z}}$$

The admittance of line 11 may be obtained from Equation (6.58) and is

$$Z_z = \frac{Z_0 \mu_r \Delta x \Delta y}{\Delta L \Delta z}$$

Hence, a matching or absorbing boundary condition for the HSCN implies a reflection coefficient [25] of

$$\rho = \frac{Z - Z_z}{Z + Z_z} = \frac{\Delta L - \sqrt{\mu_r \varepsilon_e}\, \Delta x}{\Delta L + \sqrt{\mu_r \varepsilon_e}\, \Delta x} \tag{6.65}$$

6.5 AN ALTERNATIVE DERIVATION OF SCATTERING PROPERTIES

An elegant method for obtaining the scattering properties of a TLM node, which at the same time leads to an efficient computational algorithm, is described in Refs. [27] and [28]. It is based on enforcing continuity of the electric and magnetic fields and conservation of charge and magnetic flux. It illustrates clearly the structure and symmetries of the scattering process. To best implement this method, it is convenient to rename the voltage pulses at each port using one superscript and three subscripts. The superscript is either i or r to indicate incident and reflected pulses. The first subscript indicates the direction of propagation (x, y, z). The second subscript is either n or p to indicate a line segment along the negative or positive axis. Finally, the third subscript, which can be x, y, or z, indicates the direction of polarization of the voltage pulse. Voltage pulses on ports 1 through 12, labeled using this notation, are shown in Fig. 6.19, along with the port numbers of the original notation. The alternative method is applied to study the scattering properties of the HSCN with open circuit stubs and electric and magnetic losses. Stub voltage pulses are indicated with a superscript (i or r) and two subscripts. The first subscript is either o (open-circuit), e (electric loss), or h (magnetic loss). The second subscript

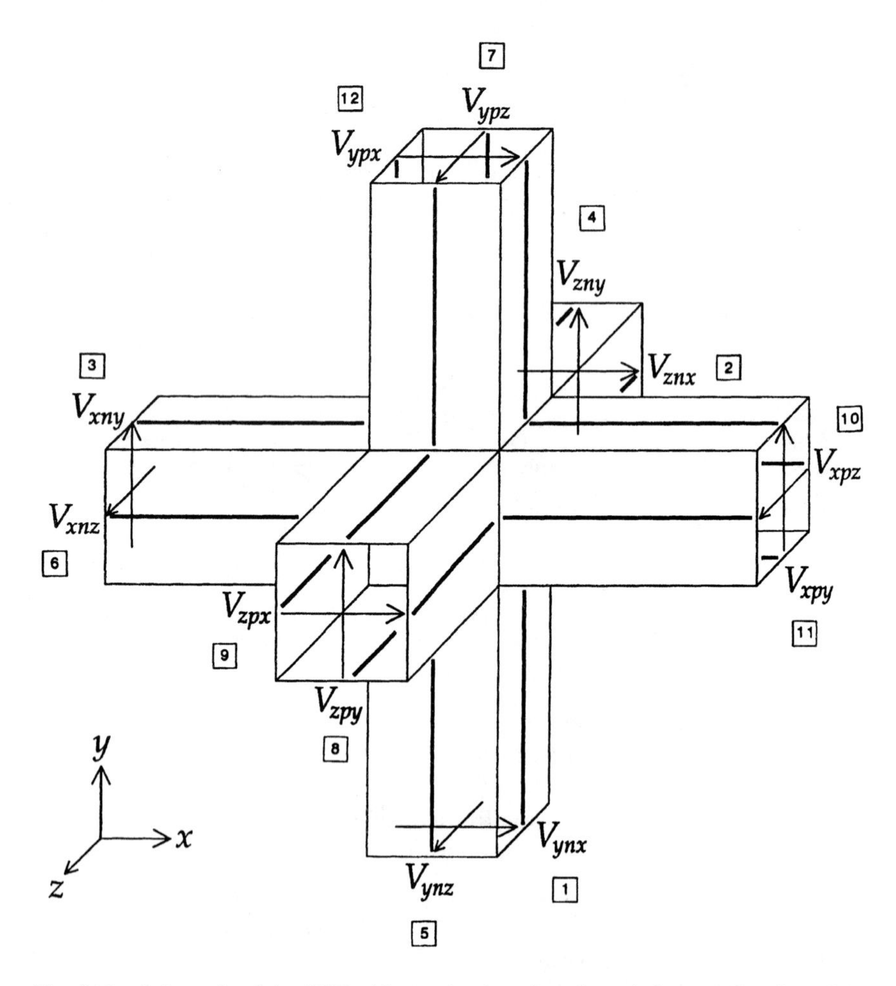

Fig. 6.19 Schematic of the SCN with notation based on the polarity and direction of the lines

is x, y, or z to indicate the polarization direction. The normalized (to Y_0) link line admittances are $\hat{Y}_x$, $\hat{Y}_y$, $\hat{Y}_z$. Similarly, the normalized stub admittances are

$$\hat{Y}_x^s,\ \hat{Y}_y^s,\ \hat{Y}_z^s \text{ for the open-circuit stubs}$$

$$\hat{G}_x,\ \hat{G}_y,\ \hat{G}_z \text{ for the electric loss stubs}$$

The normalized impedances (to Z_0) for the magnetic loss stubs are $\hat{R}_x$, $\hat{R}_y$, $\hat{R}_z$. The scattering process is determined by adhering to the following four principles explained in the following paragraphs.

Principle 1: Charge Conservation

This principle is illustrated by applying it to all the lines contributing to E_x, as shown in Fig. 6.20a.

$$\begin{aligned} &\hat{Y}_y Y_o\left[\left(V^i_{znx} - V^r_{znx}\right) + \left(V^i_{zpx} - V^r_{zpx}\right)\right] \\ &+ \hat{Y}_z Y_o\left[\left(V^i_{ynx} - V^r_{ynx}\right) + \left(V^i_{ypx} - V^r_{ypx}\right)\right] \\ &+ \hat{Y}^s_x Y_o\left(V^i_{ox} - V^r_{ox}\right) - \hat{G}_x Y_o V^r_{ex} = 0 \end{aligned} \tag{6.66}$$

This equation expresses the requirement that the charge associated with the incident pulses be equal to the charge associated with the reflected pulses.

The capacitance associated with each of the y-directed link lines may be obtained from Equation (5.29) and is $\hat{Y}_z\, Y_o\, \Delta t/2$. Similarly, for each of the z-directed lines, the capacitance is $\hat{Y}_y\, Y_o\, \Delta t/2$ and for the open-circuit stub is $\hat{Y}^s_x\, Y_o\, \Delta t/2$. The x-directed equivalent voltage V_x thus may be obtained from the expression

$$V_x \sum_i C_i = \sum_i V_i C_i \tag{6.67}$$

or, substituting for C_i,

$$\begin{aligned} &V_x \frac{Y_o \Delta t}{2}\left[2\hat{Y}_y + 2\hat{Y}_z + 2\hat{Y}^s_x\right] \\ &= (V_{znx} + V_{zpx})\,\hat{Y}_y Y_o \Delta t/2 + (V_{ynx} + V_{ypx})\,\hat{Y}_z Y_o \Delta t/2 + V_{ox}\hat{Y}^s_x Y_o \Delta t/2 \end{aligned} \tag{6.68}$$

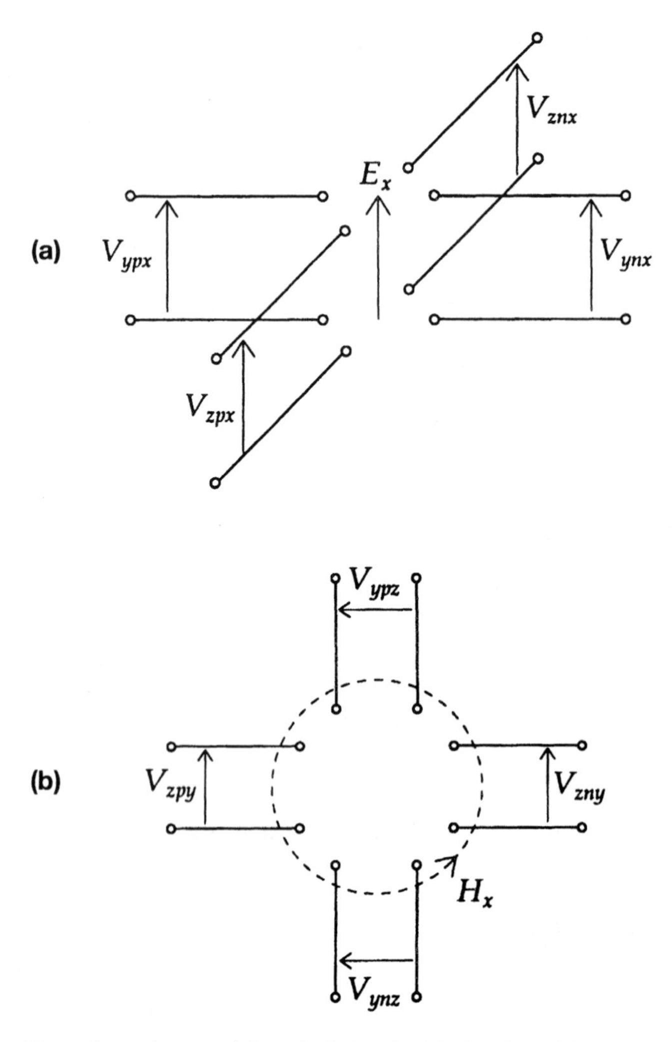

Fig. 6.20 Network topology used for calculating the (a) electric and (b) magnetic field in the HSCN (from Ref. [28])

Expressing each of the voltage terms in Equation (6.68) in terms of the sum of the incident and reflected voltage pulses, using Equation (6.66) to express the reflected voltage pulses in terms of the incident pulses, and exploiting the fact that $V^r_{ox} = V_x - V^i_{ox}$, $V^r_{ex} = V_x$, gives the following expression for the x-directed equivalent voltage:

$$V_x = \frac{2\hat{Y}_y\left(V^i_{znx} + V^i_{zpx}\right) + 2\hat{Y}_z\left(V^i_{ynx} + V^i_{ypx}\right) + 2\hat{Y}^s_x V^i_{ox}}{2\hat{Y}_y + 2\hat{Y}_z + \hat{Y}^s_x + \hat{G}_x} \tag{6.69}$$

Similar expressions are obtained for the *y*- and *z*- directed equivalent voltages V_y and V_z respectively.

Principle 2: Magnetic Flux Conservation

For *n* lines crossing a closed surface, flux conservation may be expressed in the form

$$\sum_n L_n \left(I_n^i + I_n^r \right) = 0$$

where the current pulses I_n^i and I_n^r are incident and reflected along each line. Applying this principle to the lines shown in Fig. 6.20b, contributing to the *x*-directed fields gives

$$\frac{\Delta t}{2\hat{Y}_x Y_o}\left[\left(I_{ynz}^i + I_{ynz}^r\right) + \left(I_{zpy}^i + I_{zpy}^r\right) - \left(I_{ypz}^i + I_{ypz}^r\right) - \left(I_{zny}^i + I_{zny}^r\right)\right]$$

$$+ \frac{\hat{R}_x Z_o \Delta t}{2} V_{hx}^r = 0$$

(6.70)

where use of Equation (5.29) to calculate the inductance of each line has been made. Expressing current in terms of voltage pulses in Equation (6.70) and rearranging gives

$$V_{ynz}^r + V_{zpy}^r - V_{ypz}^r - V_{zny}^r + V_{hx}^r = -\left(V_{ynz}^i + V_{zpy}^i - V_{ypz}^i - V_{zny}^i \right) \quad (6.71)$$

The equivalent current, I_x, responsible for the *x*-directed component of the magnetic field, may be obtained from the expression

$$I_x \sum_i L_i = \sum_i I_i L_i \quad (6.72)$$

or, substituting for L_i and I_i,

$$I_x \frac{\Delta t \, Z_o}{2} 4 \hat{Z}_x = \frac{\Delta t \, Z_o}{2} \hat{Z}_x (I_{ynz} + I_{zpy} - I_{ypz} - I_{zny}) \quad (6.73)$$

Expressing the current terms in Equation (6.73) in terms of incident and reflected voltages, using Equation (6.71) to eliminate the reflected voltages, exploiting the fact that $V_{hx}^r = \hat{R}_x Z_0 I_x$, and rearranging gives

$$I_x = \frac{2\left(V^i_{ynx} + V^i_{zpy} - V^i_{ypz} - V^i_{zny}\right)}{Z_o\,(\,4\hat{Z}_x + \hat{R}_x)} \tag{6.74}$$

Similar expressions apply for the y- and z-directed equivalent currents I_y and I_z, respectively. In deriving Equations (6.69) and (6.74), the view has been taken that open-circuit stubs contribute to the capacitance because, after time Δt, all energy stored is associated with the electric field. Lossy stubs do not contribute to capacitance or inductance since, at all times, voltage and current are in phase, thus representing energy dissipation.

Principle 3: Continuity of the Electric Field

The symmetry of the node implies that, for the structure shown in Fig. 6.20a, the field calculated from the y-directed lines is the same as that obtained from the z-directed lines; i.e.,

$$V_{ynx} + V_{ypx} = V_{znx} + V_{zpx} \tag{6.75}$$

or, by expressing in terms of the incident and reflected voltages and rearranging

$$\left(V^i_{ynx} + V^i_{ypx}\right) - \left(V^i_{znx} + V^i_{zpx}\right) = -\left(V^r_{ynx} + V^r_{ypx}\right) + \left(V^r_{znx} + V^r_{zpx}\right) \tag{6.76}$$

Similar expressions are obtained by enforcing continuity for lines contributing to E_y and E_z.

Principle 4: Continuity of the Magnetic Field

Demanding that the magnetic field obtained from the y- and z-directed lines in Fig. 6.20b be the same gives

$$I_{ynz} - I_{ypz} = I_{zpy} - I_{zny} \tag{6.77}$$

or, expressing in terms of incident and reflected voltages and rearranging,

$$\left(V^i_{ynz} - V^i_{ypz}\right) - \left(V^i_{zpy} - V^i_{zny}\right) = \left(V^r_{ynz} - V^r_{ypz}\right) - \left(V^r_{zpy} - V^r_{zny}\right) \tag{6.78}$$

Similar expressions are obtained by enforcing continuity for lines contributing to H_y and H_z.

The scattering process is described entirely by Equations (6.69), (6.74), (6.76), (6.78), and their equivalents for the y- and z-components. The voltage pulses scattered onto the stubs are calculated directly as already indicated, and those scattered into the link lines are obtained from expressions of the type

$$V^{r}_{ynx} = V_x - Z_o I_z - V^{i}_{ypx} \tag{6.79}$$

An efficient procedure for computation is first to obtain the total voltages (V_x, V_y, V_z) and currents (I_x, I_y, I_z) from the incident voltage pulses and then use Equation (6.79) and its equivalents combined with the expressions for the stubs to obtain the scattered voltages [27, 29, 30].

A similar algorithm may be applied to the standard and stub-loaded SCN [30]. In the case of the most general stub-loaded SCN, including electric and magnetic losses, scattering requires 54 additions/subtractions and 12 multiplications. The algorithm just described is therefore more efficient than that described in Ref. [15], in which 66 additions/subtractions, 6 multiplications, and 12 divisions by 4 are required.

6.6 THE MULTIGRID TLM MESH

The principles of maintaining synchronism and connectivity in the modeling of different parts of space in TLM have been essential prerequisites in developing the different types of TLM mesh described in the previous sections. An alternative approach is to develop a TLM mesh with several regions where discretization in space and time is such as to violate these principles. Clearly, this cannot be done without a host of problems as were outlined in Section 5.2.1. The objective is to have, in the same mesh, fine and coarse regions as shown in Fig. 6.21. The meshing of space shown in this figure, which will be referred to as *multigrid TLM* (MTLM), should be contrasted with the variable meshing shown in Fig. 6.13. It is seen by comparing these two figures that the fine region cannot be completely localized with a variable mesh, while with MTLM fine regions may be described in any part of space. Synchronism is not completely lost

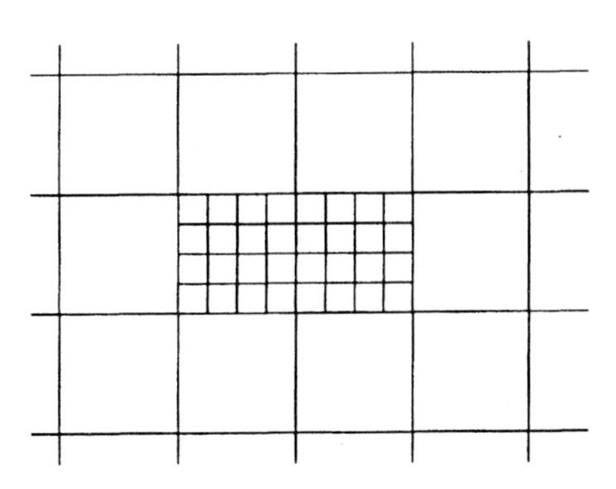

Fig. 6.21 Schematic of a multigrid TLM mesh

in MTLM, as the time-step in the coarse mesh is an integer multiple of its value in the fine mesh. The development of MTLM was described in Refs. [31] and [32]. Successful implementation of MTLM hinges on the correct processing of pulses originating from either side of a fine-coarse mesh interface. Ideally, this processing must be subject to the following constraints [32, 33]:

1. charge conservation
2. energy conservation
3. no reflections
4. zero delays

The first two requirements follow from the fact that, in a uniform mesh describing uniform space, charge and energy remain unchanged. The third requirement is necessary to make the interface between areas described by a different mesh transparent, as far as nodes adjacent to the interface are concerned. Finally, no delays must be introduced in transferring information across the interface; otherwise, incorrect resonances will be introduced. Strict adherence to all four of these requirements is not possible. The modeler is therefore forced to seek compromise conversion schemes for the pulses arriving on either side of the interface. The conversion problem may be described by reference to Fig. 6.22, showing conditions across the interface between two regions, for one polarization only, with a space resolution ratio of 2:1. In (1) the coarse region, the space and time-steps are Δl and Δt, respectively, while in (2) the fine region, there are four lines connected at the interface with space and time-steps $\Delta l/2$ and $\Delta t/2$, respectively. For the same volume of space $(\Delta l)^3$ and time period Δt, there is one pulse in region (1) and eight pulses in region (2). The conversion scheme must be capable, ideally, of converting one coarse pulse into eight fine pulses, and vice versa, while ensuring that requirements 1

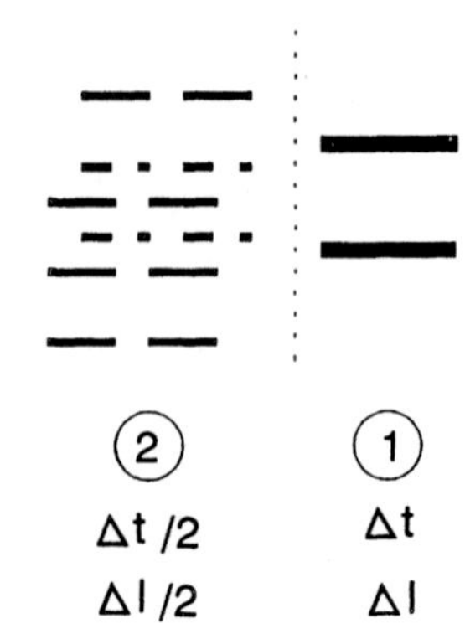

Fig. 6.22 Interface between (1) a coarse and (2) a fine mesh

through 4 are met. Extensive studies have shown that compromise conversion schemes that maintain computational simplicity and acceptable mesh performance are possible. A scheme that appears to offer such a compromise is one in which energy conservation is not explicitly enforced [33]. In practice, any resulting loss of energy is concentrated mainly at high frequencies above the cutoff frequency of the coarse mesh. It is therefore of no practical significance, except inasmuch as mesh cutoff is not a sharp one, and care must be taken when frequencies just below cutoff are considered. This is particularly important in lossless systems where repeated passes of signals across the fine-coarse interface can be expected. A way to minimize energy loss is to reduce, as far as possible, the area of the interface between fine and coarse mesh regions. Although other more rigorous conversion schemes are possible, the scheme described below has been found to give acceptable performance in a range of applications.

For pulses in region 1, the following notation is adopted: The number of lines along and across the direction of polarization is designated by s_i and p_i, respectively. The number of time-steps before conversion is t_i, and the total number of pulses before conversion is $n_i = s_i p_i t_i$. In converting pulses from region (2) to region (1), the new pulses to be inserted into region (1) are

$$V'_{1j} = \frac{1}{s_1 p_2 t_2} \sum_{k=1}^{n_2} V_{2k}, \text{ for } j = 1, 2, \ldots n_1 \tag{6.80}$$

where V_{2k} are the pulses in region (2) traveling toward the interface. The inserted pulses V'_{1j} are all taken to be of the same value to avoid introducing high-frequency components. Similarly, the new pulses to be inserted into region (2) are

$$V'_{2j} = \frac{1}{s_2 p_1 t_1} \sum_{k=1}^{n_1} V_{1k}, \text{ for } j = 1, 2, \ldots n_2 \tag{6.81}$$

where V_{1k} are the pulses in region (1) traveling toward the interface. For the arrangement shown in Fig. 6.22, the following conditions apply:

Coarse mesh (1)

$s_1 = 1, p_1 = 1, t_1 = 1, n_1 = 1$

Fine mesh (2)

$s_2 = 2, p_2 = 2, t_2 = 2, t_2 = 2 \times 2 \times 2 = 8$

Substituting into Eq. (6.80) gives

$$V'_1 = \frac{1}{4}\sum_{k=1}^{8} V_{2k}$$

This expression shows how eight fine pulses are converted into a single pulse and inserted into the coarse mesh. Similarly, substituting in Equation (6.81) gives

$$V'_{2j} = \frac{1}{2}V_1, \text{ for } j = 1, 2, \ldots 8$$

This expression shows how a single pulse, V_1, is converted into eight pulses to be inserted into the fine mesh. A schematic diagram showing the calculation order at the interface is shown in Fig. 6.23. In each mesh, pulses are inserted from the other side of the interface. Thereafter follows scattering and connection, and then pulses are removed for insertion into nodes on the other side of the interface.

This process may be modified to deal with multigrid problems using the hybrid node (HSCN). Complications arise from the fact that, in this node, three different values of impedance are used—one each for the set of lines contributing to each of the three magnetic field components. This may introduce impedance discontinuities that, however, may be dealt with as described in detail in Ref. [33]. An example of a computation involving various meshing techniques for the canonical problem proposed by ACES

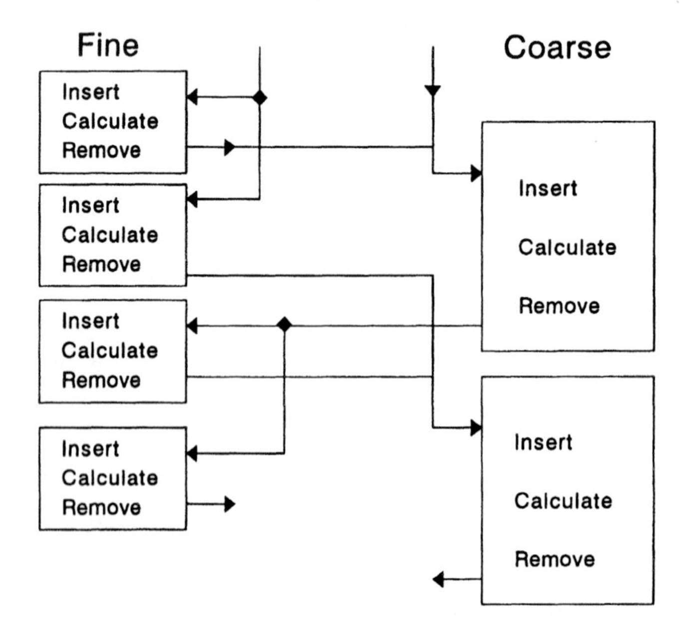

Fig. 6.23 Calculation procedure at the interface between fine and coarse mesh

[34] is shown in Fig. 6.24. Measurements of the current induced at the bottom of the coaxial cable near the ground plane, for a signal of 1 volt across 50 Ω applied to the cable, are available and are compared with simulated results obtained by using TLM. Aspects of the problem that challenge the modeler are the existence of fine features (e.g., fine wires) and the open boundary nature of the experimental configuration. It is thus inevitable that a multigrid or graded technique must be used to model as accurately as possible the entire problem space. The mesh size, storage, and run-time requirements on an HP 710 workstation for a number of model configurations are shown in Table 6.1 [27].

As a reference, results for a uniform mesh (row 1, Table 6.1) are shown by a solid line in Fig. 6.25. Although the resolution is very coarse (10 cm), the general behavior of the system is correctly predicted. Results where a finer resolution is used to describe better the wire and the table are shown for a multigrid (Fig. 6.25a) and a graded (Fig. 6.25b) mesh. For the multigrid mesh, two resolution ratios were used: 2:1 (row 2 of Table 6.1) and 4:1 (row 3 of Table 6.1). For the graded HSCN, a maximum grading of 5:1 and 10:1 was used (rows 5 and 6 of Table 6.1, respectively). Finally, a multigrid mesh, where an additional very fine resolution (8:1) near the top of the wire was employed, was used to compare with experimental results as shown in Fig. 6.26. The computational requirements for this configuration are shown in row 4 of Table 6.1. It can be seen from these simulations that more accurate results may be obtained by increasing resolution, and that the multigrid mesh offers significant economies in the use of computational resources.

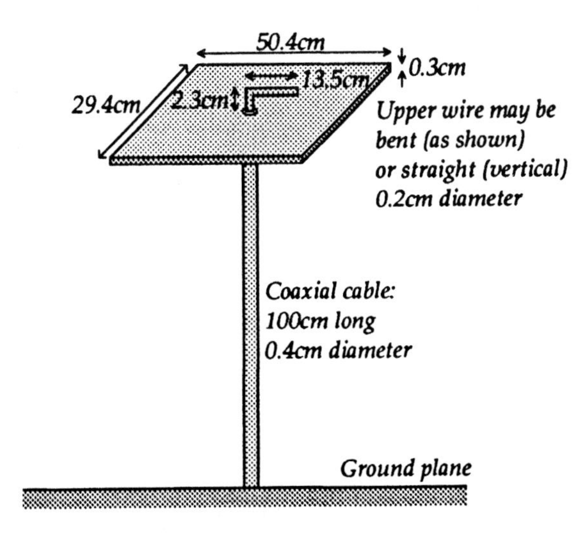

Fig. 6.24 Canonical problem for comparing simulations with experiment (from Ref. [28])

Table 6.1 Computing requirements for examples shown in Figures 6.25 and 6.26 (from Ref. [27])

	Mesh	*Storage (Mbyte)*	*Run time (min)*
1	$90 \times 90 \times 45$ (10 cm)	16.8	77
2	$90 \times 90 \times 45$ (10 cm), $8 \times 8 \times 18 + 20 \times 20 \times 5$ (5 cm)	18.4	93
3	$90 \times 90 \times 45$ (10 cm), $16 \times 16 \times 36 + 40 \times 40 \times 16$ (2.5 cm)	19.9	119
4	$90 \times 90 \times 45$ (10 cm), $16 \times 16 \times 36 + 40 \times 40 \times 16$ (2.5 cm), $32 \times 8 \times 16$ (1.25 cm)	20.2	123
5	$81 \times 81 \times 57$ hybrid node, graded $\Delta l_{min} = 2$ cm	30.2	806
6	$81 \times 81 \times 57$ hybrid node, graded $\Delta l_{min} = 1$ cm	30.2	1612

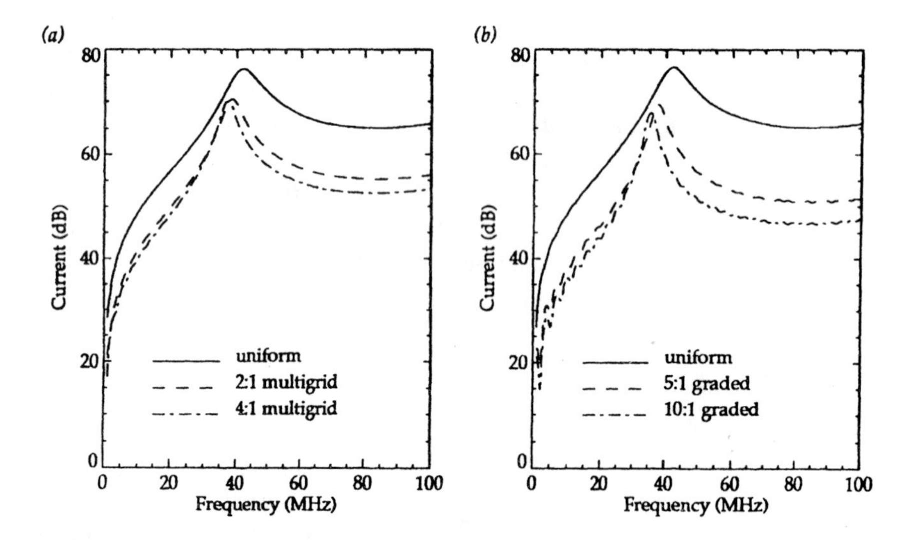

Fig. 6.25 Current induced on a wire modeled using (a) multigrid and (b) graded mesh (from Ref. [27])

REFERENCES

[1] Akhtarzad, S., and P.B. Johns. 1975. Solution of Maxwell's equations in three space dimensions and time by the TLM method of numerical analysis. *Proceedings of IEE* 122, 1344–1348.

[2] Akhtarzad, S. 1975. Analysis of lossy microwave structures and microstrip resonators by the TLM method. Ph.D. Thesis. University of Nottingham, England.

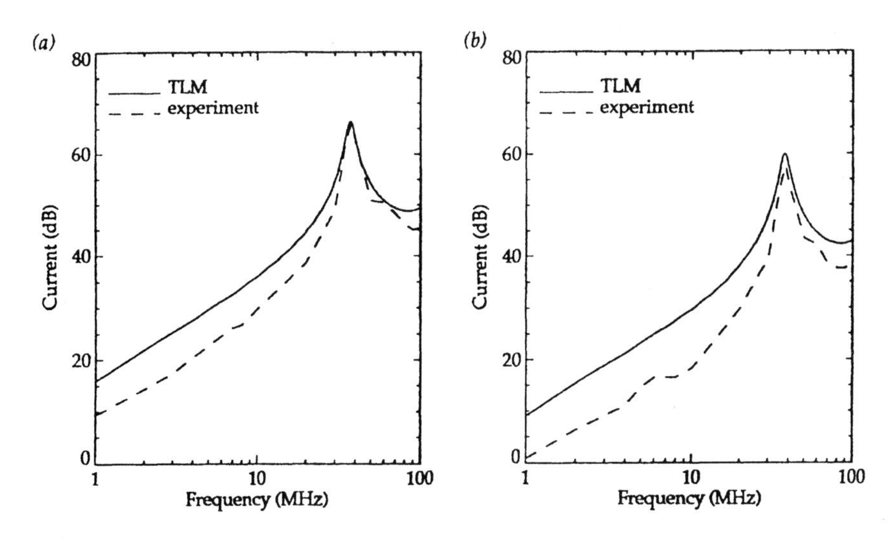

Fig. 6.26 Current induced on the wire: comparison of simulation with measurement for (a) a straight wire and (b) a bent wire (from Ref. [27])

[3] Akhtarzad, S., and P.B. Johns. 1975. Three-dimensional transmission-line matrix computer analysis of microstrip resonators. *IEEE Transactions* MTT-23, 990–997.

[4] Saguet, P., and E. Pic. 1981. Le maillage rectangulaire et le changement de maille dans la methode TLM en deux dimensions. *Electronics Letters* 17, 277–278.

[5] Al-Mukhtar, D.A. and J.E. Sitch. 1981. Transmission-line matrix method and irregularly graded space. *Proceedings of IEE* Pt H 128, 299–305.

[6] Hoefer, W.J.R. 1985. The transmission-line matrix method—theory and applications. *IEEE Transactions* MTT-33, 882–893.

[7] Saguet, P., and E. Pic. 1982. Utilisation d'un nouveau type de noeud dans la methode TLM en 3 dimensions. *Electronics Letters* 18, 478–480.

[8] Amer, A. 1980. The condensed node TLM method and its application to transmission in power systems. Ph.D. Thesis, University of Nottingham, England.

[9] Johns, P.B. 1987. A symmetrical condensed node for the TLM method. *IEEE Transactions* MTT-35, 370–377.

[10] Johns, P.B. 1986. New symmetrical condensed node for three-dimensional solution of electromagnetic wave problems by TLM. *Electronics Letters* 22, 162–164.

[11] Collin, R.E. 1966. *Foundations for Microwave Engineering.* New York, McGraw Hill.

[12] Allen, R.A, A. Mallik, and P.B. Johns. 1987. Numerical results for the symmetrical condensed TLM node. *IEEE Transactions* MTT-35, 378–382.

[13] Choi, C.H. 1989. A comparison of the dispersion characteristics associated with the TLM and FD-TD methods. *International Journal of Numerical Modelling* 2, 201–214.

[14] Nielsen, J., and W.J.R. Hoefer. 1991. A complete dispersion analysis of the condensed node TLM mesh. *IEEE Transactions on Magnetics* 27, 3982–3985.

[15] Tong, C.E.T., and Y. and Fujino. 1991. An efficient algorithm for transmission line matrix analysis of electromagnetic problems using the symmetrical condensed node. *IEEE Transactions* MTT-39, 1420–1424.

[16] Naylor, P., and R.A. Desai. 1990. New three dimensional symmetrical condensed node for solution of EM wave problems by TLM. *Electronics Letters* 26, 492–494.

[17] German, F.J., G.K. Gothard, and L.S. Riggs. 1990. Modelling of materials with electric and magnetic losses with the symmetrical condensed TLM method. *Electronics Letters* 26, 1307–1308.

[18] Dawson, J.F. 1993. Improved magnetic losses for TLM. *Electronics Letters* 29, 467–468.

[19] Allen, R., and M.J. Clark. 1988. Application of the symmetrical transmission-line matrix method to the cold modelling of magnetrons. *International Journal of Numerical Modelling* 1, 61–70.

[20] Meliani, H., D. De Cogan, and P.B. Johns. 1988. The use of orthogonal curvilinear meshes in TLM models. *International Journal of Numerical Modelling* 1, 221–238.

[21] Ward, D.D. and C. Christopoulos. 1989. A three-dimensional model of the lightning channel. *Proceedings of the 1989 International Conference on Lightning and Static Electricity,* Bath, UK, 26–28 Sept., 6B3.1–6B3.6.

[22] Wright, S. 1988. The application of transmission-line modelling implicit and hybrid algorithms to electromagnetic problems. Ph.D. Thesis, University of Nottingham, England.

[23] Voelker, R.H., and R.J. Lomax. 1990. A finite-difference transmission line matrix method incorporating a non-linear device model. *IEEE Transactions* MTT-38, 302–312.

[24] Scaramuzza, R.A. and A.J. Lowery. 1990. Hybrid symmetrical condensed node for the TLM method. *Electronics Letters* 26, 1947–1949.

[25] Scaramuzza, R.A., and C. Christopoulos. 1992. Developments in transmission line modelling and its application in electromagnetic field simulation. *International Journal for Computation and Mathematics in Electrical and Electronic Engineering* 11, 49–52.

[26] Scarammuzza, R.A., P.N. Naylor, and C. Christopoulos. 1991. Numerical simulation of field-to-wire coupling using transmission line modelling. International Conference on Computation in Electromagnetics, 25–27 November. London: IEE. *Conf Publ 350*, pp 63–66.

[27] Herring, J.L., and C. Christopoulos. 1993. The application of different meshing techniques to EMC problems. *Proceedings of the 9th Annual*

Review of Progress in Applied Computational Electromagnetics, Mar 22–26, Monterey, Calif., 755–762.

[28] Herring, J.L. 1993. Developments in the Transmission-Line Modelling Method for Electromagnetic Compatibility Studies. Ph.D. Thesis, University of Nottingham, England.

[29] Naylor, P., and R. Ait-Sadi. 1992. Simple method for determining 3D nodal scattering in nonscalar models. *Electronics Letters* 28, 2353–2354.

[30] Trenkic, V., C. Christopoulos, and T.M. Benson. 1993. Simple and elegant formulation of scattering in TLM nodes.*Electronics Letters 29,* 1651–1652.

[31] Herring, J.L., and C. Christopoulos. 1991. Multigrid transmission-line modelling method for solving electromagnetic field problems. *Electronics Letters* 27, 1794–1795.

[32] Christopoulos, C., and J.L. Herring. 1992. Developments in the transmission-line modelling (TLM) method. *Proceedings of the 8th Annual Review of Progress in Applied Computational Electromagnetics,* March 16–20, Monterey Calif., 523–530.

[33] Herring, J.L., and C. Christopoulos. 1994. Solving electromagnetic field problems using a multiple grid transmission-line modelling method. *IEEE Transactions on AP* (in publication).

[34] Hubing, T.H. 1990. Calculating the currents induced on wires attached to opposite sides of a thin plate. (See Fig. 1 in Special Publication of the Applied Computational Electromagnetics Society, *ACES Collection of Canonical Problems,* Set 1, ed. H.A. Sabbagh, 9–13.

7

The Application of TLM to Diffusion Problems

In Section 1.3, the relationship between circuit models and wave propagation in a lossy medium was described in general terms. It was shown that, with a suitable selection of circuit components, any situation ranging between lossless wave propagation and diffusion can be described. In previous chapters, modeling of problems that can be described by a wave equation with losses was presented. In this Chapter, attention is focused on diffusion problems and on thermal diffusion in particular. The form of the thermal diffusion equation is

$$\frac{\partial \theta}{\partial t} = \frac{k_{th}}{S} \nabla^2 \theta + \frac{Q}{S} \tag{7.1}$$

where

θ = temperature (a function of space coordinates and time)

k_{th} = the thermal conductivity (in W/Km)

S = the specific heat (in J/Km^3)

Q = the heat source term (in W/m^3)

TLM models for problems described by Equation (7.1) are sought in one, two, and three dimensions [1].

7.1 ONE-DIMENSIONAL DIFFUSION MODELS

The complete set of thermal equations in one dimension is

$$J = -k_{th}\frac{\partial\theta}{\partial x} \tag{7.2}$$

$$-\frac{\partial J}{\partial x} = S\frac{\partial\theta}{\partial t} - Q \tag{7.3}$$

$$\frac{\partial\theta}{\partial t} = \frac{k_{th}}{S}\frac{\partial^2\theta}{\partial x^2} + \frac{Q}{S} \tag{7.4}$$

where J is the heat flux (in W/m^2).

The circuit model described in Section (4.1) and shown in Fig. 4.1 may be adapted by setting $G = 0$ and adding a current source I to produce the circuit shown in Fig. 7.1. Kirchhoff's voltage and current laws for this circuit give

$$\Delta x\frac{\partial v}{\partial x} = -L\frac{\partial i}{\partial t} - iR \tag{7.5}$$

$$\Delta x\frac{\partial i}{\partial x} = -C\frac{\partial v}{\partial t} + I \tag{7.6}$$

These may be combined in the usual way to give

$$\frac{\partial^2 v}{\partial x^2} = \frac{LC}{(\Delta x)^2}\frac{\partial^2 v}{\partial t^2} + \frac{RC}{(\Delta x)^2}\frac{\partial v}{\partial t} - \frac{R}{(\Delta x)^2}I$$

It will be assumed that the first term on the right-hand side of this equation can be made negligible by making L very small. Under these circumstances, the equation above reduces to

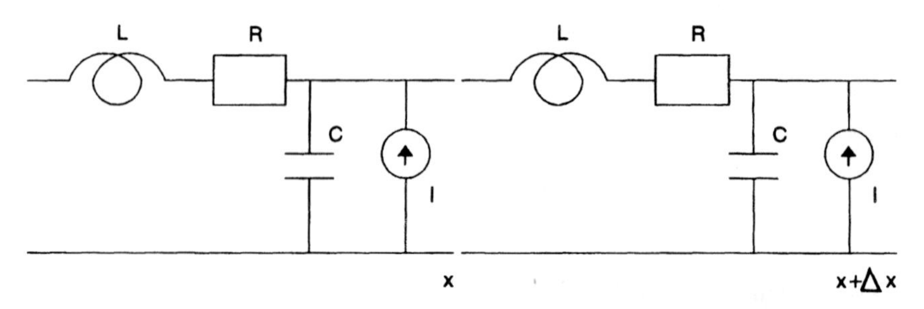

Fig. 7.1 Lumped circuit used to model 1D thermal conduction

$$\frac{\partial v}{\partial t} = \frac{(\Delta x)^2}{RC}\frac{\partial^2 v}{\partial x^2} + \frac{I}{C} \tag{7.7}$$

Equation (7.5) to (7.7) and (7.2) to (7.4) are isomorphic with the following equivalence

$$V \leftrightarrow \theta$$

$$i \leftrightarrow J\,\mathrm{A}$$

$$R \leftrightarrow \frac{\Delta x}{K_{th} A}$$

$$C \leftrightarrow S\,\Delta x\,A$$

$$I \leftrightarrow Q\,\Delta x\,A \tag{7.8}$$

where A is the cross-sectional area transverse to the direction of propagation x.

The circuit shown in Fig. 7.1 can be used as a model for heat diffusion, provided that $L = 0$. Using a link line to model the remaining capacitance in this circuit results in the TLM model shown in Fig. 7.2. In this circuit, no inductance has been explicitly included. If Δt is the transit time in the link lines, and therefore the time-step of the calculation, the impedance of the lines is then $Z = \Delta t/\mathrm{C}$. Associated with the link model of the capacitor is an error inductance which may be obtained from Equation (3.2) and is $L = (\Delta t)^2/\mathrm{C}$. For the model shown in Fig. 7.2 to be a good representation of the diffusion equation, it must be true that

$$\left|\frac{LC}{(\Delta x)^2}\frac{\partial^2 v}{\partial t^2}\right| \ll \left|\frac{RC}{(\Delta x)^2}\frac{\partial v}{\partial t}\right| \tag{7.9}$$

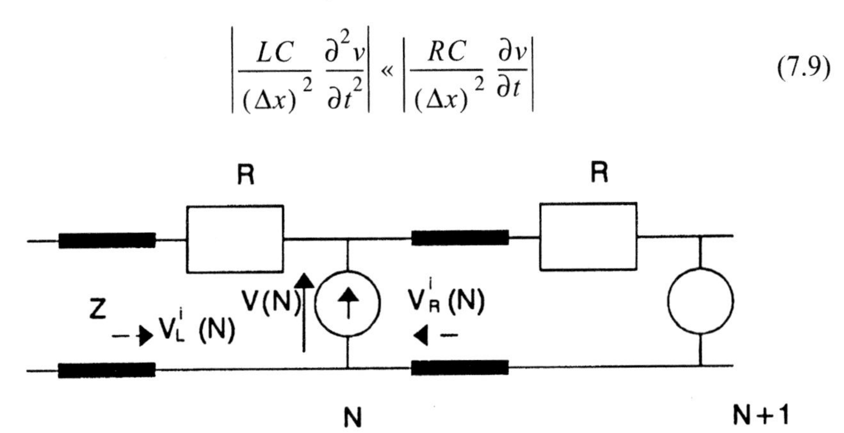

Fig. 7.2 TLM equivalent of the circuit shown in Fig. 7.1

or, following the approach described in Section 1.3, that

$$\omega L \ll R$$

Substituting L in this expression gives

$$\omega \ll \frac{RC}{(\Delta t)^2}$$

Substituting the thermal equivalents of R and C results in the condition

$$\omega \ll \left(\frac{\Delta x}{\Delta t}\right)^2 \frac{S}{k_{th}} \tag{7.10}$$

Inequality (7.10) gives the range of frequencies for which the TLM model is consistent with the diffusion equation. Additional restrictions apply as regards accuracy as for any other TLM model. Expression (7.10) may be used to establish a suitable relationship between space and time steps, depending on the range of frequencies of interest.

Computation proceeds in a similar manner as for any other TLM model. Initial conditions and source terms are defined, followed by scattering and connection in the model. At node N shown in Fig. 7.2, the voltages $V(N)$ and $V'(N)$ may be found from the Thevenin equivalent shown in Fig. 7.3. These correspond to the temperature in the thermal problem. The reflected voltages are found from

$$V_L^r(N) = V'(N) - V_L^i(N)$$

$$V_R^r(N) = V(N) - V_R^i(N)$$

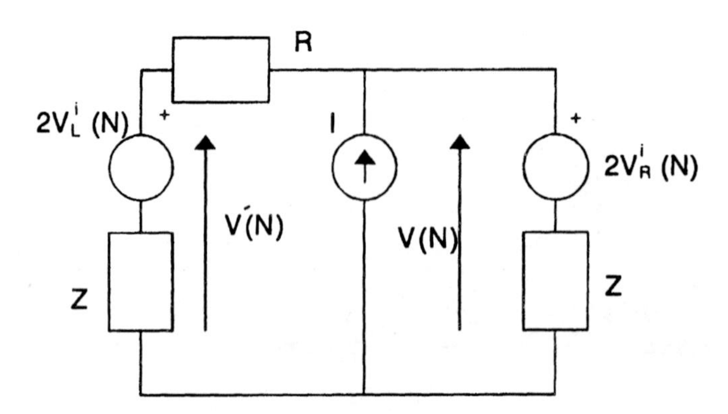

Fig. 7.3 Thevenin equivalent for conditions at node N of the circuit shown in Fig. 7.2

The new incident voltages to node N are found from the reflected voltages from the nodes connected to it. It is evident that the TLM model results in an explicit algorithm—calculation at each point depending only on local conditions during the previous time step, without the need to solve simultaneous equations. Furthermore, the TLM model is unconditionally stable. This means that the time step may be increased considerably above the value that would be allowed with an explicit finite-difference algorithm [1, 2]. This, however, cannot be done without an increase in errors, especially in cases where rapid signal variations are expected. An efficient way of implementing the algorithm is to use a time-step that varies according to conditions [3]. It should be noted that the explicit nature of the algorithm is due to the wave-like term (second derivative of v with respect to time), which however must be kept small. A discussion relating to the significance of this term may be found in Refs. [4] and [5].

7.2 TWO-DIMENSIONAL DIFFUSION MODELS

In two dimensions, the operator ∇^2 in (7.1) is $\partial^2/\partial x^2 + \partial^2/\partial y^2$. A model structure suitable for a TLM model is shown in Fig. 7.4. Adopting the approach described in Ref. [1], the following equations are obtained:

From (KVL) in x-direction

$$\frac{\partial v}{\partial x} = -L_d \frac{\partial i_x}{\partial t} - 2R_d\, i_x \tag{7.11}$$

From (KVL) in y-direction

$$\frac{\partial v}{\partial y} = -L_d \frac{\partial i_y}{\partial t} - 2R_d\, i_x \tag{7.12a}$$

From (KCL) at the node

$$\frac{\partial i_x}{\partial x} + \frac{\partial i_y}{\partial y} = -2C_d \frac{\partial v}{\partial t} + \frac{I}{\Delta \ell} \tag{7.12b}$$

where it has been assumed that $\Delta x = \Delta y = \Delta \ell$.

Differentiating Equations (7.11) and (7.12a) with respect to x and y (respectively), adding the resulting equations, and combining with Equation (7.12b) to eliminate current dependence gives the following equation:

$$\frac{\partial^2 v}{\partial x^2} + \frac{\partial^2 v}{\partial y^2} = 2L_d\, C_d \frac{\partial^2 v}{\partial t^2} + 4R_d\, C_d \frac{\partial v}{\partial t} - 2R_d \frac{I}{\Delta \ell}$$

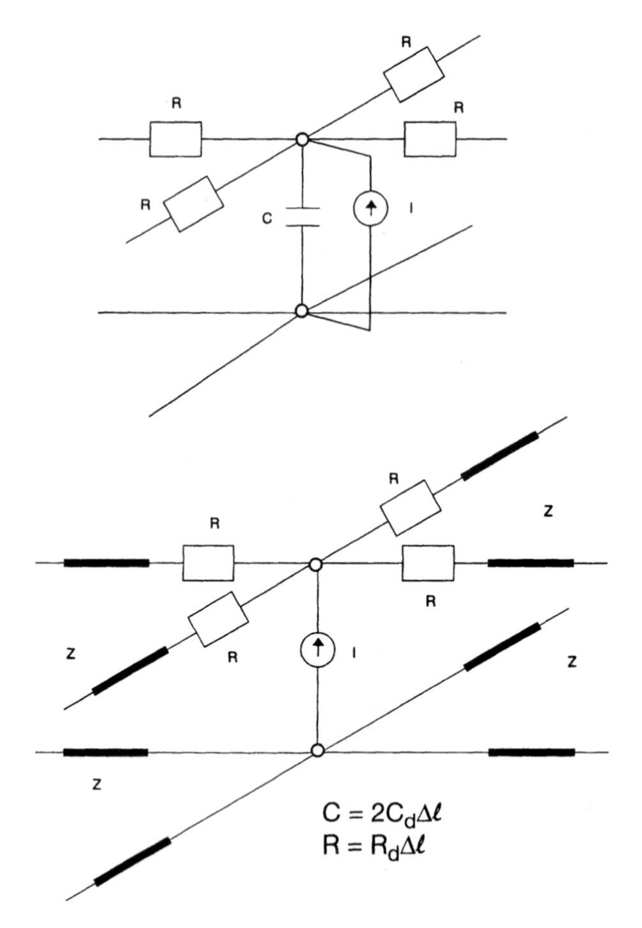

Fig. 7.4 2D node for thermal problems shown with (a) lumped and (b) TLM representation

It is assumed that the model is operated under conditions for which the second derivative term is negligible. Under these circumstances, the equation above is reduced to

$$\frac{\partial v}{\partial t} = \frac{1}{4R_d C_d}\left(\frac{\partial^2 v}{\partial x^2} + \frac{\partial^2 v}{\partial y^2}\right) + \frac{I}{2C_d \Delta\ell} \tag{7.13}$$

Comparing this with Equation (7.1), and using the same reasoning as for the 1D model, establishes the following equivalences:

$$v \leftrightarrow \theta$$

$$i \leftrightarrow J\,A$$

$$R_d \Delta \ell \leftrightarrow \frac{\Delta \ell / 2}{k_{th} A}$$

$$2 C_d \Delta \ell \leftrightarrow S\ A\ \Delta \ell$$

$$I \leftrightarrow Q\ \Delta \ell\ A \tag{7.14}$$

where $A = \Delta \ell x$ (thickness of block of material in the z-direction).

The impedance of each line is $Z = \Delta t/(C/2)$. The associated error inductance is then

$$L_d \Delta \ell = \frac{(\Delta t)^2}{C/2} = \frac{(\Delta t)^2}{C_d \Delta \ell}$$

For the wave-like term (second derivative) to be negligible,

$$2 L_d\ C_d\ \omega^2 \ll 4\ R_d\ C_d\ \omega$$

or, after substitution,

$$\omega \ll \frac{2 R_d}{L_d}$$

But $2\ R_d / L_d = (\Delta \ell / \Delta t)^2\ 2 R_d\ C_d$, hence, combining with Equation (7.14) gives the following frequency range of validity of the model:

$$\omega \ll \left(\frac{\Delta \ell}{\Delta t} \right)^2 \frac{S}{2\ k_{th}} \tag{7.15}$$

7.3 THREE-DIMENSIONAL DIFFUSION MODELS

A node structure suitable for three-dimensional diffusion is shown in Fig. 7.5. (KVL) in the x- and y-directions is as shown in Equations (7.11) and (7.12). In addition, for the z-direction,

$$\frac{\partial v}{\partial z} = - L_d \frac{\partial i_z}{\partial t} - 2\ R_d\ i_z \tag{7.16}$$

(KCL) at the node gives

$$\frac{\partial i_x}{\partial x} + \frac{\partial i_y}{\partial y} + \frac{\partial i_z}{\partial z} = - 3\ C_d \frac{\partial v}{\partial t} + \frac{I}{\Delta \ell} \tag{7.17}$$

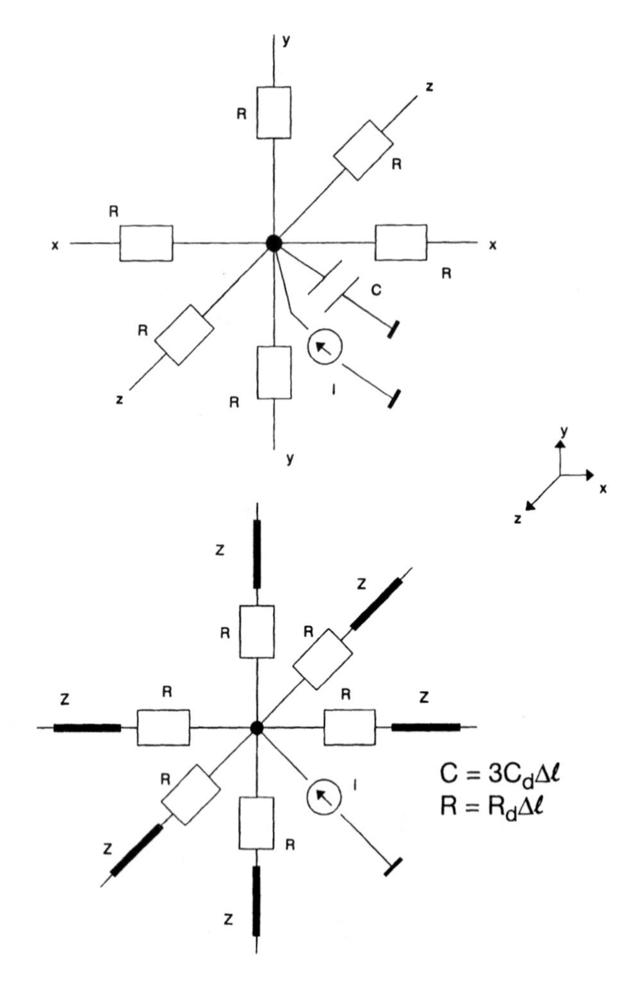

Fig. 7.5 3D node for thermal problems in (a) lumped and (b) TLM representation. All voltages are with respect to a common reference.

Differentiating (7.11), (7.12), and (7.16) with respect to x, y, z (respectively) and combining with Equation (7.17) gives

$$\nabla^2 v = 3\, L_d\, C_d\, \frac{\partial^2 v}{\partial t^2} + 6\, R_d\, C_d\, \frac{\partial v}{\partial t} - 2\, R_d\, \frac{I}{\Delta \ell}$$

Under circumstances to be determined later, the second derivative term is negligible; hence

$$\frac{\partial v}{\partial t} = \frac{1}{6\, R_d\, C_d}\, \nabla^2 v + \frac{I}{3\, C_d\, \Delta \ell} \tag{7.18}$$

Equivalences between the circuit and thermal models are then

$$v \leftrightarrow \theta$$

$$i \leftrightarrow J(\Delta\ell)^2$$

$$R_d \Delta\ell \leftrightarrow \frac{\Delta\ell/2}{k_{th}(\Delta\ell)^2}$$

$$3C_d \Delta\ell \leftrightarrow S\ (\Delta\ell)^3$$

$$I \leftrightarrow Q\ (\Delta\ell)^3 \qquad (7.19)$$

The second derivative term is negligible, provided that

$$\omega \ll 2\ R_d\ C_d \left(\frac{\Delta\ell}{\Delta t}\right)^2$$

The impedance of each line in the three-dimensional node is $Z = \Delta t/(C/3)$, where $C = 3C_d\,\Delta\ell$.

7.4 APPLICATIONS OF THE TLM MODEL OF DIFFUSION PROCESSES

The models described in the previous sections have been applied to several engineering problems. Enhancements have been introduced such as variable meshing [6] and multigrid [7–9] techniques. Heat transfer in a moving fluid has been described in Ref. [10], and heat and mass transfer in foodstuffs in [11]. Problems in electromagnetic heating whereby an electromagnetic and a thermal combined simulation is necessary have been described in Refs. [12] and [13].

REFERENCES

[1] Johns, P.B. 1977. A simple explicit and unconditionally stable numerical routine for the solution of the diffusion equation. *International Journal of Numerical Methods in Engineering* 11, 1307–1328.

[2] Johns, P.B., and G. Butler. 1983. The consistency and accuracy of the TLM method for diffusion and its relationship to existing methods. *International Journal of Numerical Methods in Engineering* 19, 1549–1554.

[3] Pulko, S.H., et al. 1990. Automatic time-stepping in TLM routines for the modelling of thermal diffusion processes. *International Journal of Numerical Modelling* 3, 127–136.

[4] Ait-Sadi, R., and P. Naylor. 1991. Validity of TLM modelling of diffusion. *Electronics Letters* 27, 2216–2217.

[5] Gui, X., et al. 1992. An error parameter in TLM diffusion modelling. *International Journal of Numerical Modelling* 5, 129–137.

[6] Pulko, S., A. Mallik, and P.B. Johns. 1986. Application of transmission-line modelling (TLM) to thermal diffusion in bodies of complex geometry. *International Journal of Numerical Modelling* 23, 2303–2312.

[7] Wong, C.C., and S.W. Wong. 1980. Multigrid TLM for diffusion problems. *International Journal of Numerical Modelling* 2, 103–111.

[8] Ait-Sadi, R., A.J. Lowery, and B. Tuck. 1990. Combined fine-coarse mesh transmission-line modelling method for diffusion problems. *International Journal of Numerical Modelling* 3, 111–126.

[9] Pulko, S,H., I.A. Halleron, and C.P. Phizacklea. 1990. Substructuring of space and time in TLM diffusion applications. *International Journal of Numerical Modelling* 3, 207–214.

[10] Pulko, S., W.A. Green, and P.B. Johns. 1987. An extension of the application of TLM to thermal diffusion to include non-infinite heat sources. *International Journal for Numerical Methods in Engineering* 24, 1333–1341.

[11] Johns, P.B., and S.H. Pulko. 1987. Modelling of heat and mass transfer in foodstuffs. In *Food Structure and Behaviour.* London: Academic Press.

[12] Desai, R.A., et al. Computer modelling of microwave cooking using the transmission-line model. *IEE Proc-A*, 139, 30–38.

[13] De Leo, R., G. Cerri, and P. Mariani. 1991. TLM technique in microwave ovens analysis: Numerical and experimental results. International Conference on Computation in Electromagnetics (25–27 November 1991, London), *IEE Conf. Publ. 350,* 361–364.

8

TLM in Vibration and Acoustics

It should come as no surprise that TLM models may be successfully developed and applied to study the behavior of systems involving mechanical vibrations and sound waves. The easiest way to establish these models is to demonstrate the isomorphism between the equations describing such systems and those applicable to TLM models. The principles exploited in developing such models are described below for torsional and sound waves. Other cases may be treated in a similar manner.

8.1 TORSIONAL WAVES

Torsion in shafts may be studied using the free-body diagram for a shaft segment of length dx as shown in Fig. 8.1, where T is the torque and θ is the angle of twist. From Newton's law,

$$\left(T + \frac{\partial T}{\partial x}dx\right) - T = (\rho I_p)\frac{\partial^2 \theta}{\partial t^2} \tag{8.1}$$

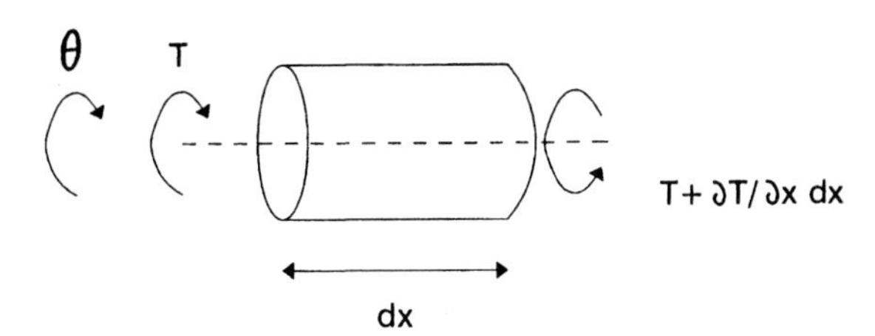

Fig. 8.1 Schematic diagram to explain the generation of torsional waves

where ρ is the density, and I_p is the polar moment of inertia. Rearranging this equation gives

$$\frac{\partial T}{\partial x} = \rho I_p \frac{\partial \omega}{\partial t} \tag{8.2}$$

where $\omega = \partial\theta/\partial t$. In addition, the following expression is obtained from torsional theory:

$$T = GI_p \frac{\partial \theta}{\partial x} \tag{8.3}$$

where G is the shear modulus of elasticity of the shaft material.

Differentiating Equation (8.3) with respect to x and combining with Equation (8.2) gives

$$\frac{\partial^2 \theta}{\partial x^2} = \frac{1}{u^2} \frac{\partial^2 \theta}{\partial t^2} \tag{8.4}$$

This is a wave equation for the angular displacement with a propagation velocity $u = \sqrt{G/\rho}$, and hence it can be modeled using the TLM method. Torque waves T^i and T^r and associated angular velocity waves ω^i and ω^r are modeled by incident and reflected voltage and current waves, respectively. Torque and angular velocity waves are related by the expression $T/\omega = \pm I_p \sqrt{\rho G}$. The expression on the right-hand side of this equation is an equivalent mechanical impedance to torsion. Boundary conditions may be applied as required. For a packed shaft ($\omega = 0$), as an example, the corresponding TLM boundary condition would be that of an open-circuit. Further details and examples may be found in Ref. [1].

8.2 SOUND WAVES

The force equation for acoustic small amplitude processes in a linear inviscid fluid is [2]

$$\rho_0 \frac{\partial \mathbf{v}}{\partial t} = -\nabla p \tag{8.5}$$

where ρ_0 is the equilibrium density of the fluid, $\mathbf{v}$ is the particle velocity, and p is the pressure. Similarly, the continuity equation is

$$\nabla \cdot \mathbf{v} = -k \frac{\partial p}{\partial t} \tag{8.6}$$

where k is the coefficient of compressibility. Combining Equations (8.5) and (8.6) gives

$$\nabla^2 p = (k\rho_0) \frac{\partial^2 p}{\partial t^2} \tag{8.7}$$

This is a wave equation for the pressure with a propagation velocity

$$u = \sqrt{\frac{1}{k\rho_0}}$$

An analogy with the TLM model may be established where pressure and particle velocity correspond to voltage and current, respectively. The density and compressibility are represented in the model by inductance and capacitance, respectively. Further details and applications may by found in Ref. [3].

REFERENCES

[1] Partridge, G. J., C. Christopoulos, and P.B. Johns. 1987. Transmission line modelling of shaft dynamic systems. *Proceedings of the Institute of Mechanical Engineers* 201, C4, 271–278.

[2] Kinsler, L.E., A.R. Frey, A.B. Coppens, and J.V. Sanders. 1982. *Fundamentals of Acoustics.* New York: John Wiley & Sons.

[3] Saleh, A.H.M., and P. Blanchfield, P. 1990. Analysis of acoustic radiation patterns of array transducers using the TLM method. *International Journal of Numerical Modelling* 3, 39–56.

9

Application of TLM to Electromagnetic Problems

In the previous chapters, the theoretical development of TLM and the basics of its application were described. Emphasis was placed on describing in some detail the physical basis of the model, starting from simple lumped and one-dimensional problems, and extending the coverage to include the most general three-dimensional inhomogeneous problems. The physical nature of TLM and the systematic model construction presented here should make it straightforward for the reader to use the model in many practical situations and also to extend and modify it to deal with problems specific to each particular application. In this chapter, further help is offered by dealing specifically with certain common applications so that the reader may become aware of the power and range of applicability of the TLM method.

9.1 ELECTROMAGNETIC COMPATIBILITY

Electromagnetic compatibility (EMC) is an aspect of equipment behavior in which the equipment is immune to certain levels of electromagnetic interference (EMI) and, at the same time, contributes a limited amount of EMI to the environment. Thus, EMC studies cover the widest possible range of frequencies and equipment. There are two aspects to EMC. First, emission of EM radiation from equipment must be controlled at the source and as it propagates through shields and interfaces. Second, the susceptibility of equipment to EM radiation that penetrates through shields, apertures, and cable interfaces must be ascertained. Legally enforced EMC standards have meant that a considerable design effort is required to ensure, in advance of final construction, that equipment meets

EMC specifications. This avoids excessive delays and costs during redesign. Electromagnetic modeling using numerical techniques such as TLM can thus assist in the design process by establishing the degree of coupling and interference between electronic systems and subsystems, and by pointing to optimum design strategies.

The modeling problems encountered in EMC studies are as follows [1]:

- The shape and configuration of equipment is very general, and the simulation models used must be capable of describing very general geometries.
- Nonuniformities (e.g., conductors, dielectrics, and so forth) are commonly encountered in large numbers in general equipment, and it is thus necessary to be able to deal with abrupt and frequent changes in material properties.
- Modeling for EMC requires fine resolution to deal with typical wire sizes, and also the capability to model open (infinite) boundary problems. These large differences in physical scale impose severe computational and modeling requirements.
- Commercial EMC extends from very low frequencies to at least 1 GHz. Special and military systems cover an even wider range. It is therefore of considerable importance to develop models that cover as simply as possible a wide frequency range.

To illustrate the application of TLM to EMC, a generic problem (namely, field-to-wire coupling) will be examined in some detail. A schematic of such a configuration is shown in Fig. 9.1. The problem is nor-

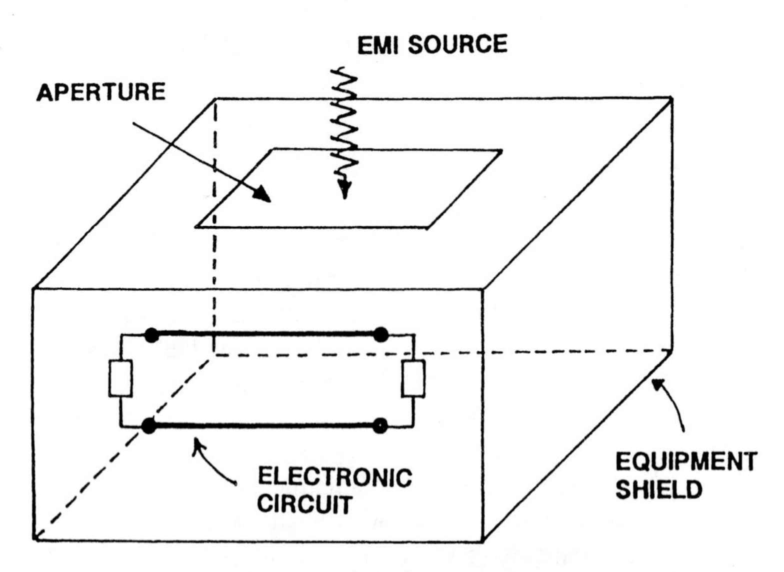

Fig. 9.1 Schematic diagram for the study of field-to-wire coupling

mally posed as follows. An electromagnetic field is incident on a conducting box with an aperture. Inside the box, there is a transmission line representing part of some important circuit. It is required to calculate the current induced on the line if the level of the electromagnetic threat (the incident wave) is known. In constructing a TLM model for this problem, a regular SCN, a hybrid SCN, or a multigrid mesh may be used. If the "wires" are large in diameter or the "loom" current is required, as in a bundle of wires, it may be possible to model the entire space using a regular mesh. Otherwise, an HSCN or a multigrid mesh will be necessary to enhance resolution around the wire(s).

Special "wire nodes" can also be used to model fine wires. This technique is explained further in Chapter 10. The problem as posed is an open boundary problem, and it requires for its numerical solution the imposition of an artificial boundary containing the box shown in Fig. 9.1. Maximum economy in the use of computer resources implies that the artificial numerical boundary is placed as close as possible to the box. The space between this boundary and the box is sometimes referred to in the literature as "white space" and must be minimized for computational efficiency. The simplest approach in TLM is to use an "absorbing" or "matched" boundary condition on the numerical boundaries. Provided enough "white space" is used, this solution is satisfactory in terms of accuracy except for the most demanding applications. Better boundary conditions in TLM have been proposed to increase accuracy and efficiency [2].

The outline of the box and apertures is described in the model by imposing short-circuit boundary conditions at suitable locations between the SCN nodes. The description of the "wire" is determined to a large extent by the size of its diameter, D, relative to the size of the nodes. If $D \geq \Delta\ell$, one or more nodes may be used with appropriate short circuits to describe the wire cross-section. This is shown in Fig. 9.2a and b, where one and 2×2 nodes, respectively, are used. Higher accuracy can be

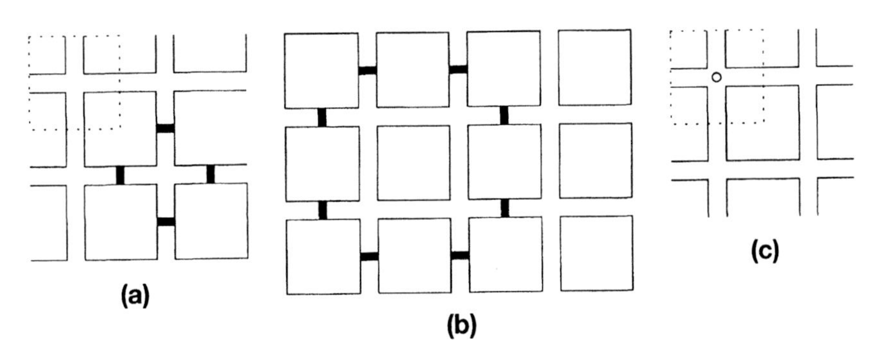

Fig. 9.2 Various methods of describing wires in the TLM mesh

achieved if a fine mesh is used around the wire by applying the multigrid technique described in Section 6.6. An alternative approach, which is the only one possible if $D << \Delta\ell$, is to employ special wire nodes where the scattering matrix is modified to account for the presence of the wire. In such models, the wire diameter may be set independently of the node size. This arrangement is shown schematically in Fig. 9.2c and is discussed in more detail in Chapter 10. Terminations may be introduced by employing lossy nodes and short circuits as required.

An example of such a calculation is shown in Fig. 9.3, where a very coarse wire description was used (single short-circuit node). The response shown is for an incident double exponential pulsed wave penetrating through the aperture and was obtained by convolving the impulse response with the actual input pulse [3]. A more complex simulation for a canonical EMC problem was discussed in Chapter 6 (Figs. 6.24–6.26).

The treatment of wires in the previous two examples is such that they form an integral part of the three-dimensional field solution. This is described as an "integrated" solution, its disadvantage being that the normally large difference between the overall problem size and the diameter of wires can create efficiency and resolution problems in modeling and computation. The use of a "wire node" as described in Chapter 10 offers an alternative approach that overcomes some of these difficulties.

An altogether different approach is to employ a "separated solution," whereby the field problem is solved assuming that wires are not present. The fields thus obtained (incident fields) are used to define equivalent voltage and current sources which are used as excitation terms in a transmission line problem consisting of the wire and its return [4]. An example of such a model for a wire plus return configuration is shown in Fig. 9.4. Multiconductor problems may also be treated in a similar manner [5, 6]. This approach is considerably simpler to apply, and it allows for the treatment of many wires closely spaced in bundles. However, it does not allow for reradiation from the wires (incorrect Q-factor of resonances), and its accuracy is limited except in simple cases where TEM propagation applies [7]. TLM has been applied to a wide range of EMC problems including lightning characterization [8], determination of induced surface currents on aircraft [9], automotive EMC [10], screened room damping [3], and calibration for EMC measurements [11].

9.2 MICROWAVE DESIGN

The characterization and design of microwave devices and circuits is based to a large extent on semi-empirical formulae and lumped circuit

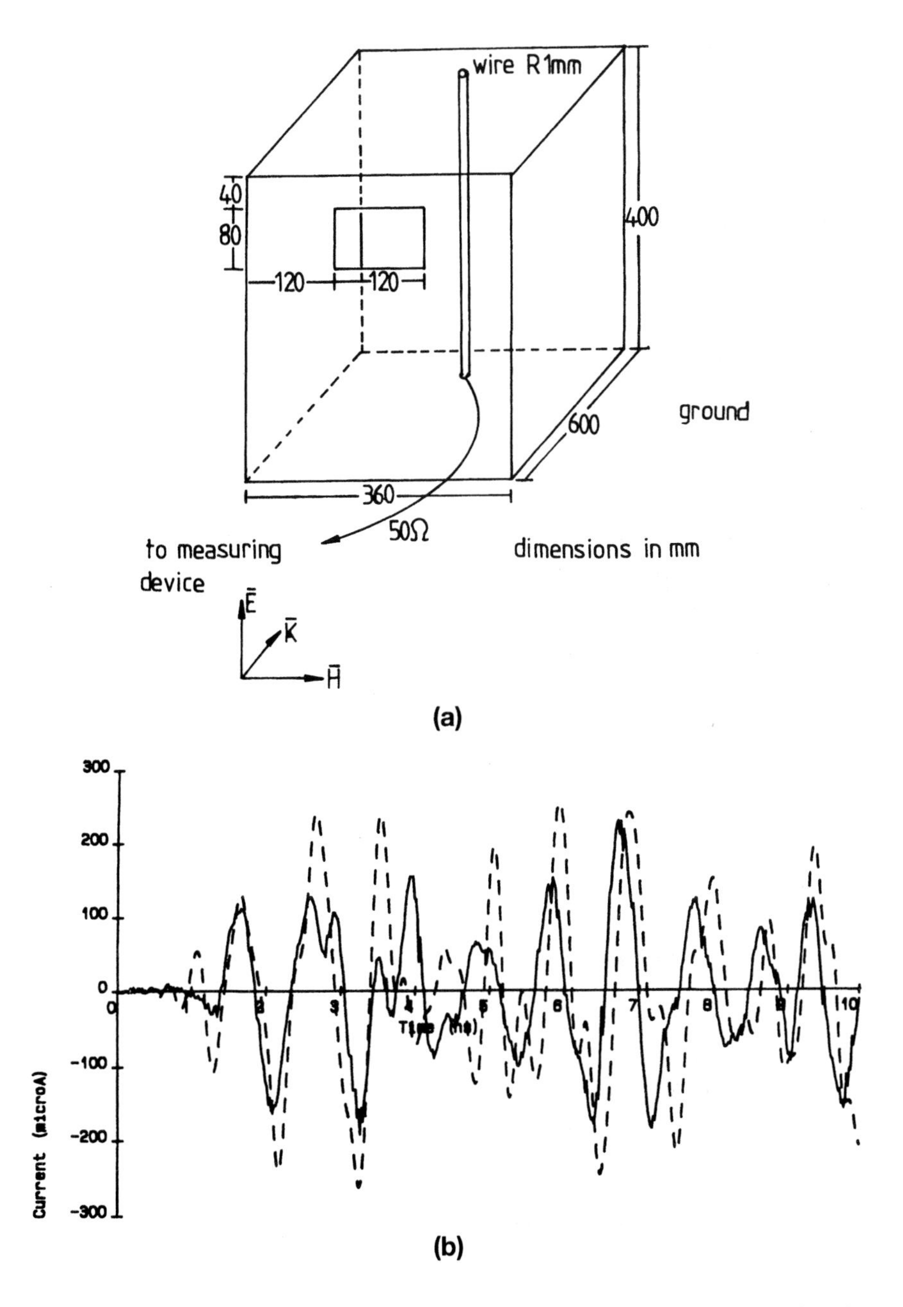

Fig. 9.3 (a) conducting box with aperture subject to an incident plane wave. The wire diameter is 2 mm. (b) Current induced on the wire, obtained from experiment (solid line) and simulation (broken line). (From "Transmission-Line Modelling in EMC Studies," by J.L. Herring et al., *International Journal of Numerical Modelling* 4. Copyright © 1991 by John Wiley & Sons., Ltd. Reprinted by permission.)

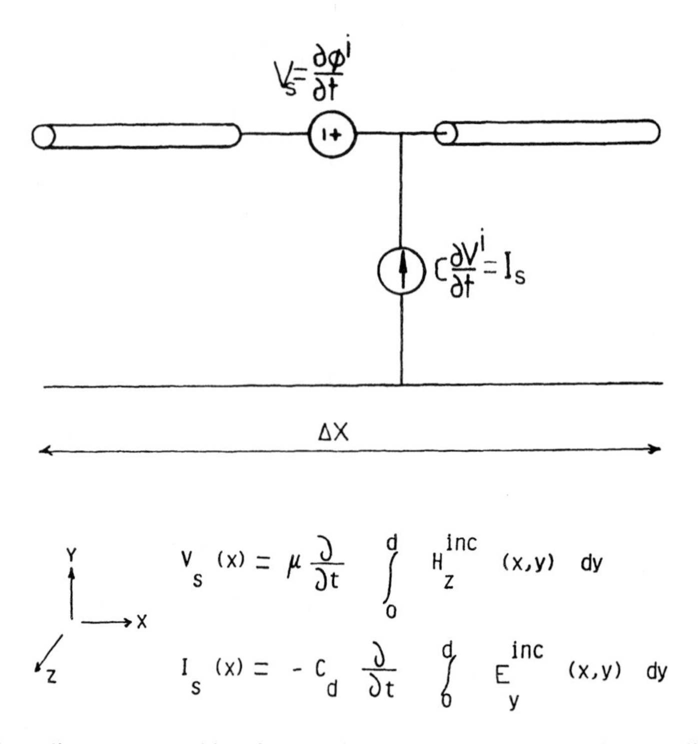

Fig. 9.4 A line segment with voltage and current sources representing coupling with incident fields

representations. The advent of powerful EM simulation software is making a growing contribution to microwave design, providing self-consistent solutions to increasingly complex problems. An example of a microwave problem where a field solution is necessary is shown in Fig. 9.5. This figure shows the junction between two strips of different width in a micro-

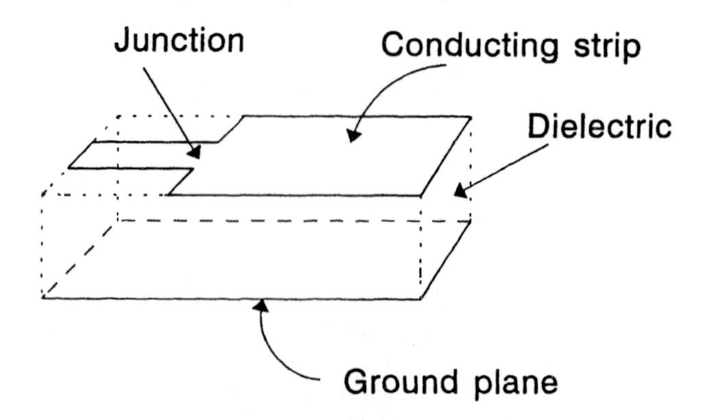

Fig. 9.5 Schematic diagram of a junction in a microstrip line

strip line. The characteristic impedance to the left and to the right of the junction may be obtained, but it would be an approximation to assume that fields are abruptly established on either side of the junction to suit the particular conditions for each transmission line. Hence, a full field solution near the junction is necessary to account for the detailed structure of the field as it encounters the junction and propagates past it. TLM has been applied to such problems from its early days [12]. The modeling principles and techniques are similar to those explained in Section 9.1. However, the configurations of interest in microwave circuits tend to be simpler and better defined so that greater accuracy can be achieved. Of particular interest in problems such as the one depicted in Fig. 9.5 is the determination of the *s*-parameters. Simulation normally proceeds by launching an impulse some distance from the junction, obtaining the reflected and transmitted pulses, and thus (eventually) the *s*-parameters. A major factor affecting the accuracy of this calculation is the elimination of (or, if this is not possible, the compensation for) signal reflections from numerical boundaries. In general, the impedance of absorbing boundaries is frequency dependent and hence cannot be easily represented in a time-domain simulation by a single component or reflection coefficient. A useful approach is to employ the so-called "Johns Matrix" which is similar to a reflection coefficient matrix but where the reflected pulses at time t are a function of incident pulses at t and all previous time intervals [13–15]. Numerical reflection problems can be minimized, in principle, if the length of the microstrip line beyond the junction included in the model is increased. However, the effects of this extra length in the simulation may be included by using the Johns Matrix and diakoptics. This is illustrated in Fig. 9.6, which shows a section to the right of line $00'$ which is to be represented by a Johns Matrix [15]. This structure interacts with the rest of the circuit to the left of $00'$, which is to be modeled in detail, through the input nodes ($n = 1$ to N). The same nodes ($m = 1$ to M) are also the

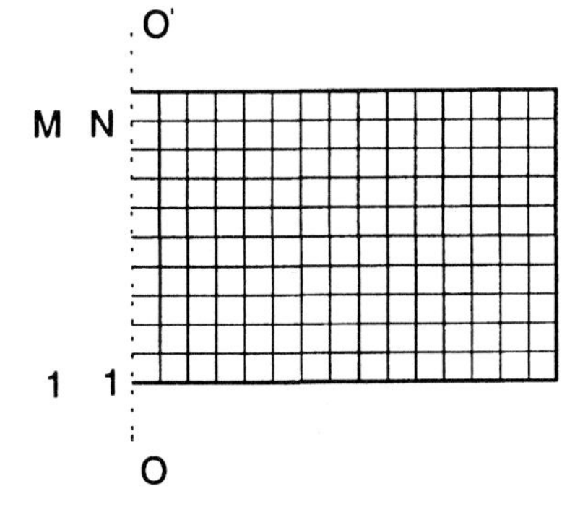

Fig. 9.6 Schematic diagram of the area to the right of $00'$ to be described by a Johns Matrix. (From "The Electromagnetic Wave Simulator," by W.J.R. Hoefer and P.P.M. So. Copyright © 1991 by John Wiley & Sons, Ltd. Reprinted by permission.)

output nodes. As far as the problem to the left of 00′ is concerned, the extended boundary is described fully if the reflected pulse at time kΔt on output port m is determined from all incident voltages arriving on all input ports $n = 1$ to N and at times $k'\Delta t \leq k\Delta t$. This can be expressed by a convolution of the input signals and the Johns Matrix; i.e.,

$$V^{r}_{(m,k)} = \sum_{n=1}^{N} \sum_{k'=0}^{K} g(m, n, k - k')\, V^{i}_{(n,k)} \tag{9.1}$$

The Johns matrix has dimension $M \times N \times (k + 1)$, and its elements can be determined as follows:

A unit voltage pulse is injected on node 1 while all other ports on 00′ are matched. The pulse streams on all M output ports are then obtained for the total time duration $K\Delta t$. These pulse streams are elements $g(m, 1, k)$ ($m = 1$ to N, and $k = 1$ to K) of the Johns matrix. Similarly, with port 2 excited, elements $g(m, 2, k)$ are obtained. In general, N such computations are required, and this can be a considerable task. However, by exploiting symmetry and the simplicity of some of the microwave circuits, the effort required can be considerably reduced. Examples of the application of TLM to microwave circuit problems may be found in the papers already mentioned in Refs. [16] through [19].

A problem that has received considerable attention in microwave work is the shift in resonances toward lower frequencies for structures containing sharp features that are modeled with limited resolution [20]. This coarseness error may be understood by reference to the schematic diagram shown in Fig. 9.7. In this configuration, nodes A and C have direct communication with the conducting surface, unlike node B, where communication is somewhat delayed. An increase in resolution reduces these errors at the expense of a longer computation. A method that

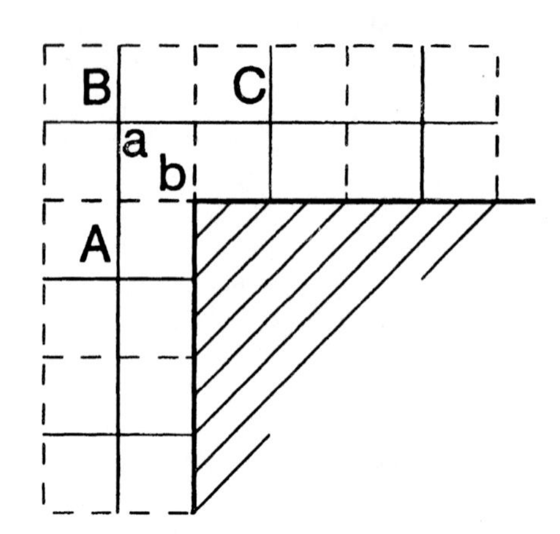

Fig. 9.7 Communication around a corner node B

improves accuracy without undue increases in computation is described in Ref. [21]. In essence, this method provides an additional signal path from *a* to *b* in node *B* of Fig. 9.8, in the form of a short-circuit stub. The parameter selection, scattering matrix, and implementation of this technique are described in detail in Ref. [21].

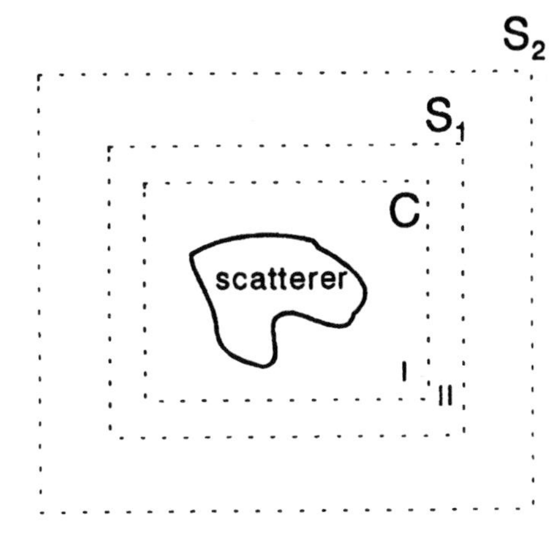

Fig. 9.8 Schematic of modeling regions used to study RCS problems

An alternative approach for dealing with this type of error, which can also be applied in three-dimensions and to any structure, is described in Refs. [22] and [23]. In this approach, the delay in communication in corner nodes is corrected by a small reduction in ε_r and μ_r from their physical value corresponding to the medium surrounding the conductor. The reduction factor is calculated in Ref. [23], where it is also demonstrated that for a wire, communication in the azimuthal direction rather than along the length of the wire is responsible for the errors. Although correction at the corner nodes is sufficient, it is computationally simpler to apply a correction on the single layer of nodes immediately adjacent to the wire, as explained in the cited reference. Such corrections may be necessary if the coarseness of the mesh and the required accuracy of computations demand it. In an uncorrected simulation, errors in wire resonances of a few percentage points should be expected. It should be emphasized that these errors are observed in conducting structures with external corners (wires, strips, etc.) and are not present in cavity resonances, which are predicted with very high accuracy.

9.3 RADAR CROSS-SECTION (RCS)

A problem of considerable interest is the determination of the scattering properties of objects subject to incident electromagnetic waves. Such problems occur in radar studies, and a common quantity to be determined is the so-called *radar cross-section* (RCS) of an object defined as

$$\sigma = 2\pi \lim_{r \to \infty} \left(\frac{r|E^s|^2}{|E^i|^2} \right) \tag{9.2}$$

where E^s and E^i are the scattered and incident fields, respectively, and r is the distance between the observation point and the scatterer [24].

The same modeling principles apply as in other applications, but special problems arise due to the nature of the RCS calculation. First, it is clear that an incident plane wave must be set up and propagated toward the object. Propagation must be with the minimum of distortion or reflections from numerical boundaries. Second, the scattered wave must be determined and must remain unaffected by reflections from numerical boundaries. Third, since Equation (9.2) implies a calculation of the scattered fields at a large distance from the scatterer, and this in turn would require a prohibitively large computation, the scattered fields must be obtained on a closed surface surrounding the scatterer and then extrapolated using a near-to-far field transformation based on the equivalence principle [25]. The manner in which such calculations proceed is shown in Fig. 9.8. The objective of such studies is to obtain the scattered field from an object subject to an incident electromagnetic field. Space around the scatterer is separated by surface C into two regions. Inside C (region I) the total fields (incident plus scattered) are calculated, while outside (region II) the scattered fields only are allowed to propagate. Typically, C runs through the middle of link lines joining adjacent nodes. Voltage pulses at C traveling from region I to II, and from region II to I, have pulses

$$V_+^i \text{ and } V_-^i$$

respectively, added to them. The latter two pulses are related to the incident field as explained in detail in Ref. [26]. The scattered fields evaluated on surface S_1 are used in conjunction with a near-to-far field transformation to determine the field at any desired distance [25, 27]. Surface S_2 marks the termination of the simulation volume, and on it a high quality absorbing boundary condition is imposed.

9.4 ANTENNAS

The TLM method may be used to model antennas of various designs. Many of the techniques described in the previous sections also apply to the modeling of antennas and will not be repeated here. Simple wire antennas are typically modeled using the HSCN and/or a multigrid mesh. Of particular significance in this case is the high spacial resolution required to model thin wires and the errors inherent in modeling the wire cross-section by a single node. TLM is especially useful when modeling antennas in complex environments. A particular example is the study of spherical dipole antennas and effects due to their proximity to a conducting corner [1]. Similarly, TLM has been applied to the study of cavity backed aperture antennas [28]. Other examples include the study of the radiation pattern and input impedance of microstrip antennas [29] and of tapered slot antennas [30].

A problem commonly encountered is the detailed modeling of excitation and antenna feeds. Wire antennas may be excited in a variety of ways. Assuming that the wire is described by short circuits placed in the middle point of link lines, incident voltage pulses may be imposed on adjacent ports as shown schematically in Fig. 9.9a. Similarly, a current may be induced by imposing a magnetic field as shown in Fig. 9.9b (Ampere's Law). Finally, a voltage source V_S of internal impedance Z_S may be connected anywhere as an antenna feed, as shown in Fig. 9.9c for a feed at $(x + 1, y, z)$.

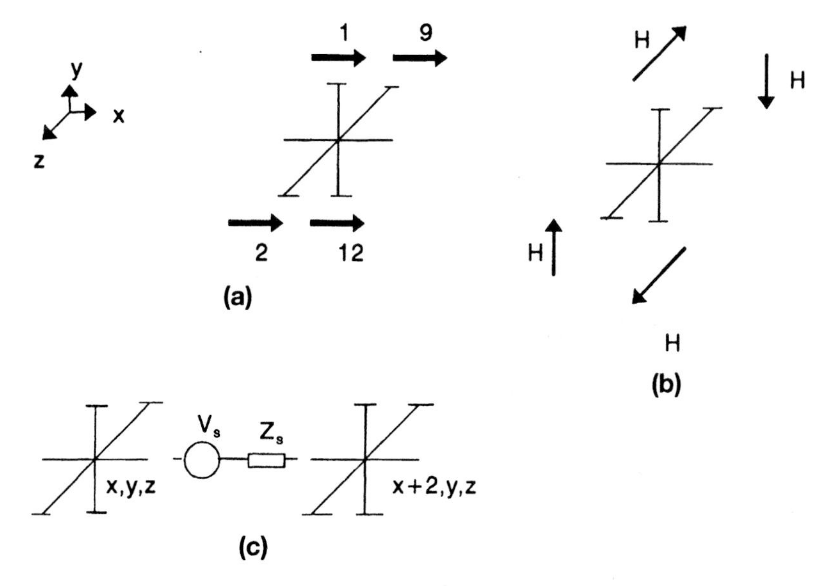

Fig. 9.9 (a) incident voltages injected on ports adjacent to a model of a wire node; (b) current injection by H-field excitation around the model of a wire; and (c) source connected to port 1 of node $(x + 1, y + 1, z)$, port 9 of node $(x + 1, y, z - 1)$, port 12 of node $(x + 1, y - 1, z)$, and port 2 of node $(x + 1, y, z + 1)$

9.5 ELECTROMAGNETIC HEATING

A particular class of problems revolves around the use of high-frequency electromagnetic waves for the heating of materials. Applications range from microwave cooking to bioelectromagnetics [31]. The power absorbed per unit volume in a material is given by the expression

$$P = \omega \, \varepsilon_0 \, \varepsilon''_{eff} \, |E|^2 \tag{9.3}$$

where $|E|$ is the root-mean-square of the electric field in the material, and ε''_{eff} is related to the complex permittivity by the expressions

$$\varepsilon = \varepsilon_0 \, (\varepsilon' - j\varepsilon'')$$

$$\varepsilon''_{eff} = \varepsilon'' + \frac{\sigma}{\omega \varepsilon_0} \tag{9.4}$$

A typical study in electromagnetic heating involves the determination of the electromagnetic field distribution using the TLM electromagnetic models described earlier, and a simultaneous thermal simulation using the techniques described in Chapter 7. Consideration must be given to the level of coupling between the electromagnetic and thermal problems. The electromagnetic model provides the source term [Equation (9.3)] for the thermal model. In addition, most parameters (e.g., ε'', thermal conductivity) are temperature dependent and, as a result, a certain amount of coupling is inevitable through these parameter variations. Strong coupling should be expected during phase changes in the material. TLM has been applied successfully to this type of problem and further details may be found in Refs. [32] through [34].

REFERENCES

[1] Christopoulos, C., and J.L. Herring. 1993. The application of transmission-line modelling (TLM) to electromagnetic compatibility problems. *IEEE Transactions* EMC-35, 185–191.

[2] Morente, J.A., J.A. Porti, and M. Khalladi. 1992. Absorbing boundary conditions for the TLM method. *IEEE Transactions* MTT-40, 2095–2099.

[3] Herring, J.L., P. Naylor, and C. Christopoulos. 1991. The application of transmission line modelling in electromagnetic compatibility studies. *International Journal of Numerical Modelling* 4, 143–152.

[4] Naylor, P., C. Christopoulos, and P.B. Johns. 1987. Coupling between electromagnetic fields and wires using TLM. *IEE Proceedings* 134, 679–686.

[5] Christopoulos, C., and P. Naylor. 1988. Coupling between electromagnetic fields and multiconductor transmission systems using TLM. *International Journal of Numerical Modelling* 1, 31–43.

[6] Naylor, P., and C. Christopoulos. 1989. Coupling between electromagnetic fields and multimode transmission systems using TLM. *International Journal of Numerical Modelling* 2, 227–240.

[7] Christopoulos, C., P. Naylor, and P.B. Johns. 1989. Numerical simulation of coupling into wires. *Proceedings of the 8th International Zurich Symposium on EMC*, 179–184.

[8] Mattos, M.A. da F., and C. Christopoulos. 1990. A model of the lightning channel, including corona, and prediction of the generated electromagnetic fields. *Journal of Phys. D.: Applied Physics* 23, 40–46.

[9] Johns, P.B., and A. Mallik. 1985. EMP response of aircraft structures using TLM. *Proceedings of the 6th International Zurich Symposium on EMC*, 387–389.

[10] Herring, J.L., and C. Christopoulos. 1990. The vehicle body as an electromagnetic shield—numerical simulation for emission and susceptibility studies. *Proceedings of the 7th International Conference on EMC*, IEE Conf. Publ. 326, (York, UK) 125–131.

[11] Herring, J.L., and C. Christopoulos. 1991. Numerical simulation for better calibration and measurements. *Proceedings of the 5th British Electromagnetic Measurements Conference* (Malvern, UK), 37/1–37/4.

[12] Akhtarzad, S., and P.B. Johns. 1975. Three-dimensional transmission-line matrix analysis of microstrip resonators. *IEEE Transactions* MTT-23 12, 990–997.

[13] Hoefer, W.J.R. 1989. The discrete time domain Green's function or Johns Matrix—A new powerful concept in transmission line modelling (TLM). *International Journal of Numerical Modelling* 2, 215–225.

[14] Eswarappa, G.I. Costache, and W.J.R. Hoefer. 1990. Transmission line matrix modelling of dispersive wide-band absorbing boundaries with time-domain diakoptics for s-parameter extraction. *IEEE Transactions* MTT-38, 379–386.

[15] Hoefer, W.J.R., and P.P.M. So. 1991. *The Electromagnetic Wave Simulator–A dynamic visual electromagnetics laboratory based on the two-dimensional TLM method.* New York: John Wiley & Sons.

[16] Mariki, G.E., and C. Yeh. 1985. Dynamic three-dimensional TLM analysis of microstrip lines on anisotropic substrate. *IEEE Transactions* MTT-33, 789–799.

[17] Saguet, P., and W.J.R. Hoefer. 1988. The modelling of multiaxial discontinuities in quasi-planar structures with the modified TLM method. *International Journal of Numerical Modelling* 1, 7–17.

[18] So, P.P.M, Eswarappa, and W.J.R. Hoefer. 1989. A two-dimensional transmission line matrix microwave field simulator using new concepts and procedures, *IEEE Transactions* MTT-37, 1877–1884.

[19] Itoh, T. (ed.). 1989. Numerical Techniques for Microwave and Millimeter-Wave Passive Structures. New York: John Wiley & Sons.
[20] Shih, Y.C., and W.J.R. Hoefer. 1980. Dominant and second-order mode cut-off frequencies with a two-dimensional TLM program. *IEEE Transactions* MTT-28, 1443–1448.
[21] Mueller, V., P.P.M. So, and W.J.R. Hoefer. 1992. The compensation of coarseness error in 2D TLM modelling of microwave structures. *Microwave Symposium Digest* MTT-S, 373–376.
[22] Duffy, A.P., T.M. Benson, C. Christopoulos, and J.L. Herring. 1993. New methods for accurate modelling of wires using TLM. *Electronics Letters,* 29, 224–226.
[23] Duffy, A.P., J.L. Herring, T.M. Benson, and C. Christopoulos. 1994. Improved wire modelling in TLM. *IEEE Transactions* MTT-42, 1978–1983.
[24] German, F.J., G.K. Gothard, L.S. Riggs, and P.M. Goggans. 1989. The calculation of radar cross-section (RCS) using the TLM method. *International Journal of Numerical Modelling* 2, 267–278.
[25] German, F.J., G.K. Gothard, and L.S. Riggs. 1990. RCS of three-dimensional scatterers using the symmetrical condensed TLM method. *Electronics Letters* 26, 673.
[26] Simons, N.R.S., A.A. Sebak, and E. Bridges. 1991. Transmission-line matrix (TLM) method for scattering problems. *Computer Physics Comms.* 68, 197–212.
[27] Yee, K.S., and K. Shlager. 1991. Time-domain extrapolation to the far field based on FDTD calculations. *IEEE Transactions* AP-39, 410–413.
[28] Duffy, A.P., T.M. Benson, and C. Christopoulos. 1993. Numerical modelling of cavity backed apertures using transmission-line modelling (TLM). *Proceedings of the 8th International Conference on Antenna and Propagation,* (Edinburgh, UK), IEE Conf. Publ. 370, 107–110.
[29] Dubard, J.L., et al. 1991. Acceleration of TLM through signal processing and parallel computing. *Proceedings of the International Conference on Computation in Electromagnetics* (London, UK), IEE Conf. Publ. 350, 71–74.
[30] Ndagijimana, F., P. Saguet, and M. Bouthinon. 1990. Tapered slot antenna analysis with 3D TLM method. *Electronics Letters* 26, 468–470.
[31] Fleming, A.H.J., and K.H. Joyner (eds). 1992. Special Issue in Bioelectromagnetic Computations. *ACES, Journal,* 7, no. 2.
[32] Desai, R.A., et al. 1992. Computer modelling of microwave cooking using the transmission-line model. *IEE Proceedings* A, 139, 30–38.
[33] De Leo, R., G. Cerci, and P. Mariani. 1991. TLM techniques in microwave oven analysis: numerical and experimental results. *Proceedings of the International Conference on Computation in Electromagnetics* (London) IEE Conf. Publ. 350, 361–364.
[34] Trenkic, V., C. Christopoulos, and J.G.P. Binner. 1993. The application of the transmission-line modelling (TLM) method in combined thermal and electromagnetic problems. *Proceedings of the International Conference on Numerical Methods for Thermal Problems* (Swansea), 1263–1274.

10

Special Topics in TLM

In previous chapters, the basic approach to TLM was systematically described, and details of some of its applications in a range of practical problems were presented. There are, however, a number of special TLM techniques that can be employed in certain problems, and these are described in this chapter. Finally, TLM implementation issues are briefly addressed to help readers with applications and problem solving.

10.1 THIN-WIRE FORMULATIONS

The modeling of thin wires is a permanent difficulty for all differential numerical modeling methods. In TLM, the cross-section of a wire may be described in detail by using a number of nodes and short circuiting the appropriate lines. This, however, is rarely a practical proposition, as computational costs for thin wires in large working volumes are excessive. Reductions in computation can be made by using the multigrid technique described in Section 6.6, but the fact remains that, in most cases, the wire cross-section has to be modeled using a single node. The way that this can be done is shown schematically in Fig 10.1a and b using a short-circuit node and shorts on link lines, respectively. In either case, the diameter of the wire is comparable to the dimensions of the node $\Delta\ell$. The exact relationship between wire radius and $\Delta\ell$ is difficult to ascertain, as it depends on the way the equivalent "electromagnetic" radius of the wire is defined. As an illustration, if the electric and magnetic fields established in a wire-above-ground configuration are determined with the wire modeled as indicated above, and the equivalent radii are calculated from the electric and magnetic field distributions, they are found to be slightly different. It

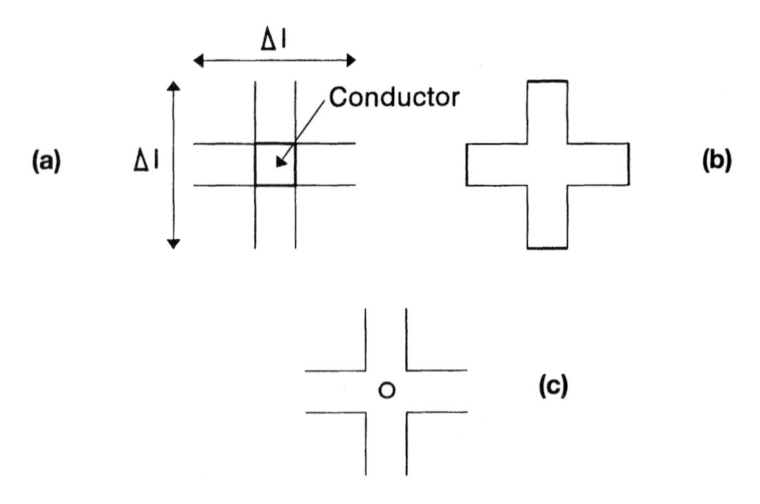

Fig. 10.1 Three ways of modeling wires: a short-circuit node (a) shorts on link lines (b) and a wire node (c)

is thus advisable to take the wire diameter as approximately equal to $\Delta\ell$ but to keep in mind the difficulties mentioned above. Modeling becomes more efficient if a way can be found to model reasonably accurately thin wires in an otherwise coarse mesh ($r << \Delta\ell$).

A thin-wire formulation in connection with the FDTD method was first described in Ref. [1]. In this approach, a wire of arbitrary radius (but smaller than the cell size) is embedded in the cell. The first attempt in developing an equivalent approach in TLM was described in Ref. [2]. The "wire node" is shown schematically in Fig. 10.1c. It consists of a standard SCN node with a wire threaded through one of the coordinate directions. Thus, in addition to the normal 12 ports of the SCN, two extra ports are introduced, representing the two ends of the wire. The admittance looking into these two ports representing the wire is calculated from standard formulae for a coaxial transmission line consisting of the wire (inner conductor) and an outer conductor of diameter $\Delta\ell$. A scattering matrix is then obtained for the new node by imposing unitary conditions as described in Ref. [2]. The determination of line admittance, based on the values of L and C obtained using the same wire radius and $\Delta\ell$ consistently as indicated above, introduces velocity errors which become apparent in some problems. Improvements have been made to this basic structure by relaxing the consistency requirement and these are described below.

The first improved wire-node structure to be developed models a thin wire placed between nodes as shown schematically in Fig. 10.2 [3] for a wire running along the z-direction and placed between nodes (x, y, z) and $(x + 1, y, z)$. The pair of ports on each node which are immediately

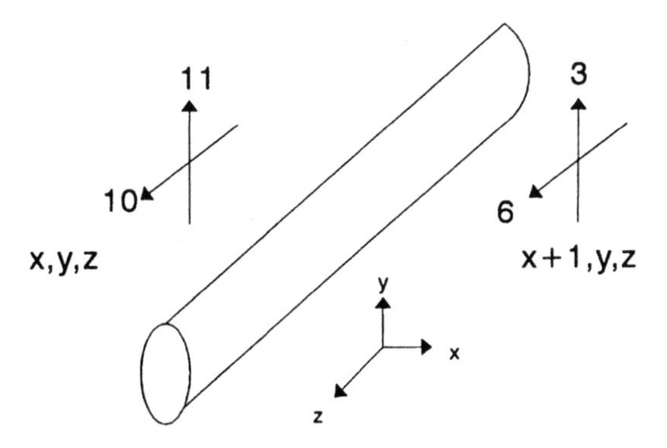

Fig. 10.2 A thin wire placed at $x + \Delta\ell/2$ running along the z-direction

adjacent to the wire are also shown. The wire is modeled by a link line of characteristic impedance Z_{line} and an inductive stub of impedance Z_{stub}. Only ports 10 of node (x, y, z) and 6 of $(x + 1, y, z)$ interact directly with the wire. The manner of interaction is shown in Fig. 10.3. The "wire" capacitance and inductance per unit length are calculated from the formulae [3]

$$C_d = \frac{2\pi\varepsilon}{\ln\left(\dfrac{0.4\Delta y}{r}\right)}, \quad L_d = \frac{\mu}{4\pi}\ln\left(\frac{0.15\Delta y}{r}\right) \tag{10.1}$$

where Δy is the node dimension in the y-direction, and factors 0.4 and 0.15 represent equivalent outer radii determined after numerical experimentation. The line parameters in the circuit shown in Fig. 10.3 can then be easily determined. Modeling all the required capacitance in lines 3 and

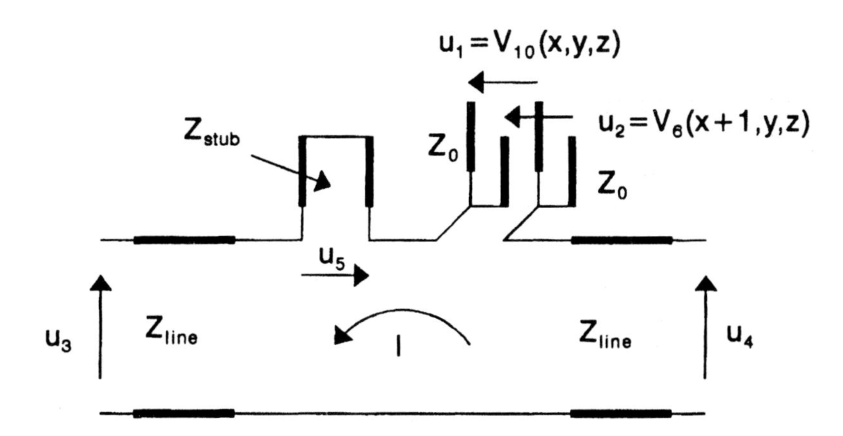

Fig. 10.3 Conditions at the interface with a thin-wire model

4 gives $Z_{line} = \Delta t / (C_d \, \Delta z)$. These two lines also model inductance 2 $Z_{line}(\Delta t/2) = (\Delta t)^2 / (C_d \, \Delta z)$. The deficit in inductance is thus

$$L_{stub} = L_d \Delta z - \frac{(\Delta t)^2}{C_d \, \Delta z}$$

Hence,

$$Z_{stub} = \frac{2L_{stub}}{\Delta t} = 2\left(\frac{L_d \, \Delta z}{\Delta t} - \frac{\Delta t}{C_d \, \Delta x} \right) \tag{10.2}$$

These choices maintain synchronism throughout the TLM mesh. Solutions are obtained by replacing all link lines and stubs in Fig. 10.3 by their Thevenin equivalents in terms of the incident voltages, calculating the current I, and thus determining the reflected voltage pulses. These pulses become the incident pulses on adjacent points at the next time-step. The relevant formulae may be found in Ref. [3]. Scattering and connection proceeds as for any SCN mesh but with the modifications at wire nodes described here. Caution must be exercised in using the formulae (10.1) for wires in close proximity to other bodies. In such cases, it may be necessary to substitute for new estimates of the empirical factors 0.4 and 0.15.

An alternative formulation, in which the wire is placed through the center of the node, is described in Ref. [4]. This method employs formulae (10.1) and requires the determination of a new scattering matrix. It is thus similar in principle to the technique described in Ref. [2]. An extension of the method to include wire loads is described in Ref. [5].

10.2 NARROW-SLOT FORMULATIONS

Narrow slots feature prominently in many problems, and their modeling presents the same difficulties as that of thin wires. An approach similar to that used for thin wires may also be used for slots by exploiting duality. A slot on an infinite conducting sheet and the dipole formed by interchanging the regions of metal and slot are complementary problems. The E-field distribution of the slot and the H-field distribution of the dipole are the same. The impedance of the dipole Z_d and of the slot Z_s are related by the expression [6]

$$Z_s Z_d = \frac{Z_0^2}{4} \tag{10.3}$$

where Z_0 is the intrinsic impedance of the surrounding medium. For a wire of radius r_1 and equivalent return radius r_2, the inductance and capacitance per unit length were given in Equation (10.1), where $r_2 = 0.4\ \Delta y$ for the capacitance, and $r_2 = 0.15\ \Delta y$ for the inductance. The slot capacitance is related to the wire inductance by Equation (10.3); i.e.,

$$\frac{1}{j\omega C_s} j\omega L_d = \frac{Z_0^2}{4}$$

Substituting for L_d and Z_0 in the above expression gives

$$C_s = \frac{2\varepsilon}{\pi}\ln\left(\frac{r_2}{r_1}\right) \tag{10.4}$$

The slot is shown in Fig. 10.4, and it follows that $r_1 = w/2$ and $r_2 = 0.563\ \Delta\ell/2$, the latter expression being established after numerical experimentation [7]. Substituting these values into Equation (10.4) and adding a contribution due to slot thickness gives the following expression for the slot capacitance per unit length:

$$C_s = \frac{2\varepsilon}{\pi}\ln\left(\frac{0.563\Delta y}{w}\right) + \frac{\varepsilon d}{w} \tag{10.5}$$

A similar procedure may be used to obtain an expression for the inductance per unit length of the slot.

$$L_s = \left[\frac{2}{\pi\mu}\ln\left(\frac{1.591\Delta y}{w}\right) + \frac{d}{\mu w}\right]^{-1} \tag{10.6}$$

where the factor 1.591 is again determined after numerical experimentation. The circuit showing the coupling between the slot model (lines 3, 4

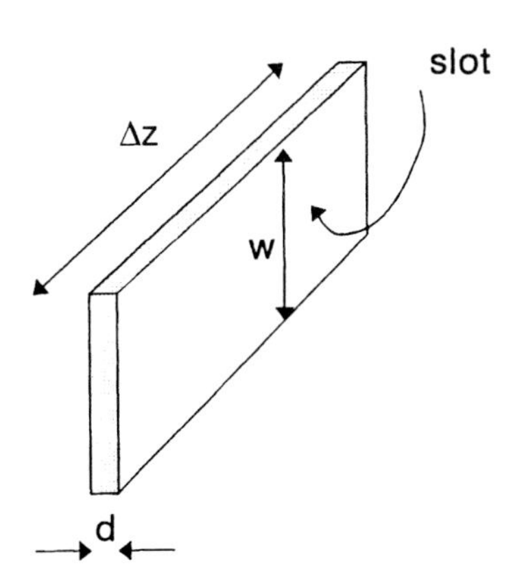

Fig. 10.4 Schematic of a narrow slot

and stub 5) and the ports of the adjacent nodes (1 and 2) is shown in Fig. 10.5. The stub for the slot is terminated by an open-circuit (capacitive stub). The slot model parameters Y_{line} and Y_{stub} are determined as for the thin wire to maintain synchronism. The voltage, V in Fig. 10.5, is then calculated from the incident voltages. The reflected and new incident voltages are then obtained, and computation proceeds in the normal way as for any other TLM mesh.

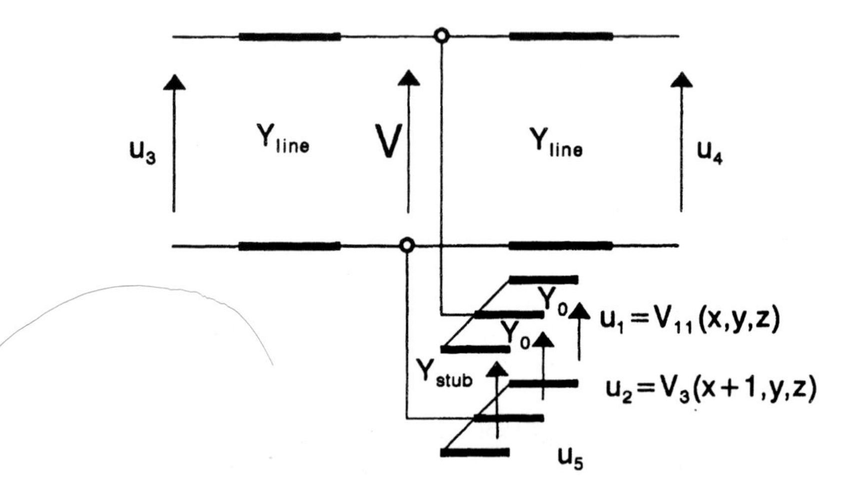

Fig. 10.5 Conditions at the interface with a narrow slot model

It is worth pointing out that, to maintain stability, all stubs must have positive impedance. Thus, in the case of the wire, it follows from Equation (10.2) that for stability,

$$r < 0.108\Delta y \tag{10.7}$$

A similar stability condition may be derived for the slot model and is

$$w < 0.398\Delta z \tag{10.8}$$

10.3 THIN-PANEL FORMULATIONS

There are many practical situations where thin panels of finite conductivity need to be described in a TLM mesh. A particular example is panels made out of carbon fiber composites (CFCs) used extensively in aircraft construction. The electrical conductivity of these panels is several orders of magnitude lower compared to the conductivity of metals. Their shielding properties are thus very different, and the EM behavior of structures containing such materials is difficult to predict. The difficulties associated with modeling a thin panel in an otherwise coarse mesh are associated with computational costs. It is simply unprofitable to attempt a detailed three-dimensional model of the thickness of the panel. Multigrid techniques can offer considerable help, but simpler, more efficient methods are necessary for most problems. One such technique is presented in Ref. [8]. In essence, it describes propagation through a thin panel as a one-dimensional event for each polarization. The panel of thickness d is divided into n layers, as shown in Fig. 10.6a. Each layer is then represented by a transmission line segment as shown in Fig. 10.6b. The capacitance, inductance, and conductance of each layer are obtained from

$$C = \frac{\varepsilon\ \Delta x\ d}{n\ \Delta y},\ L = \frac{\mu\ \Delta y\ d}{n\ \Delta x},\ G = \frac{\sigma\ \Delta x\ d}{n\ \Delta y} \tag{10.9}$$

where ε, μ, and σ are the electrical parameters of the panel. A TLM model of each circuit component is constructed using stubs as shown in Fig. 10.6c, according to the principles established in Chapter 3. Several layers may be necessary ($n \sim 10$) to give a reasonably accurate description of the panel. Using the same time-step for the stub models as for the rest of the TLM mesh maintains synchronism. Computation proceeds by injecting into the TLM model of the panel pulses coming from the field problem on the two sides of the panel. The n-layer model is solved, using standard circuit techniques, to obtain the scattered voltage pulses in terms of the two

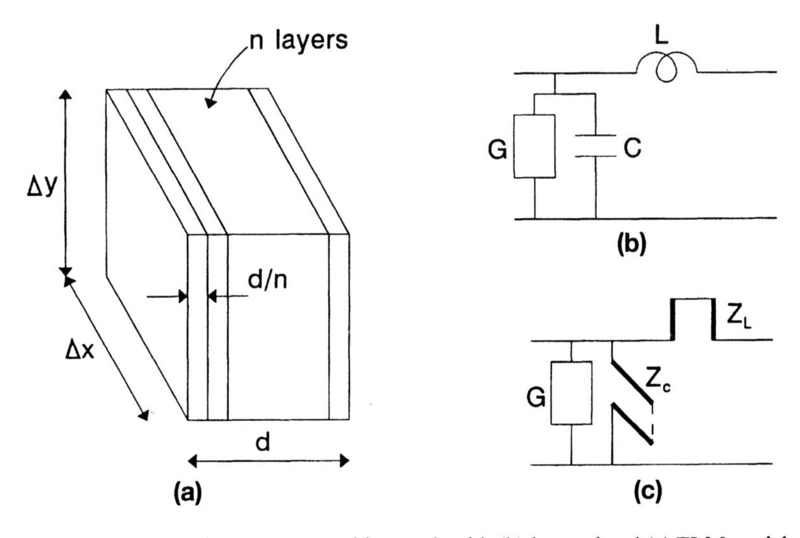

Fig. 10.6 (a) layers used to represent a thin panel, with (b) lumped and (c) TLM models of a layer

incident pulses already mentioned and all other incident pulses coming from the stubs. The scattered pulses form the new incident pulses for the model of the panel and for the TLM ports representing the field around the panel. Naturally, nonuniform panels may be easily modeled and different properties assigned to the panel for different polarizations.

10.4 INFINITELY ADJUSTABLE BOUNDARIES

In many modeling situations, where a cartesian mesh of finite spacial resolution is employed, it is necessary to use a stepwise approximation to boundaries. As an illustration, if a cylindrical boundary needs to be described in such a mesh, a staircase approximation is inevitable. A noncartesian mesh can be used as described in Section 6.3.3, but then similar problems may arise in such a mesh if a planar boundary is also present. This problem is common to all differential methods as described in Ref. [9] in connection with the FDTD method. The problem arises because, when using a standard SCN, it is only possible to accurately describe features that are placed at integer multiples of the nodal spacing $\Delta\ell$. More accurate positioning of boundaries may be achieved by using a graded and/or multigrid mesh, but the basic limitation remains, especially for boundaries making a small angle with the cartesian axes. The first attempt to remove these restrictions was reported in Ref. [10] in connection with two-dimensional TLM problems. As described earlier, normally, a boundary can only be placed a distance $\Delta\ell/2$ away from the center of a node. The approach described in Ref. [10] aims at placing the boundary at a distance $\Delta\ell/2 + \ell$. Thus, a line of length $\Delta\ell/2$ is replaced by a line of different length $(\Delta\ell/2 + \ell)$ which has the same propagation time and input impedance. This can be achieved if the longer line has a different characteristic impedance. Enforcing these requirements allows the calculation of the new characteristic impedance. A new scattering matrix is then derived, since the boundary branch of the node has a different characteristic impedance.

An alternative approach is described in Ref. [11], where the impedance of the boundary branch remains unchanged, but its termination is an inductance or a capacitance depending on whether the boundary is a short circuit (electric wall), or an open circuit (magnetic wall). The method can only be employed to lengthen the lines.

A development of the standard SCN to allow for a very general positioning of boundaries in TLM is described in Ref. [12]. In this node, the only restriction on the impedance of the link lines shown in Fig. 6.19 is that the same value must be used for both lines on each limb (e.g., $Z_1 = Z_5$). Thus, in general, there are six different values of impedance, resulting in a node where the length of each limb can be independently selected. In principle, such a node may be used to describe accurately any boundary without the staircasing feature of the standard SCN. The scattering matrix for this node is obtained after a lengthy calculation using well established principles and may be found in Ref. [12].

10.5 FREQUENCY-DOMAIN TLM (TLM-FD)

TLM has been developed and applied as a time-domain method. It is, however, possible to formulate TLM in the frequency-domain, and in this section the rudiments of this formulation are presented. It is perhaps necessary to discuss briefly why, under certain circumstances, a frequency-domain formulation may be advantageous. A general discussion of some of the relevant issues was presented in Section 1.2. More specifically, there are situations where the steady-state response is required at a single frequency, and in such cases it appears unnecessary to approach steady-state after calculating the entire transient. This would be the case if time-domain TLM is applied. The attainment of steady-state can be in some cases a very lengthy process when, for example, waves have to penetrate through panels with a very long diffusion time constant. Another example is the modeling of highly inhomogeneous materials where extensive use of stubs needs to be made, resulting in numerical dispersion and a very short time-step to maintain stability and synchronism. It is thus useful to have access to a frequency-domain code, based on the same principles as the time-domain model, which can be used when the need arises.

The first attempt to formulate TLM models using the expanded node to provide steady-state solutions to Maxwell's equations was described in Ref. [13]. The development of a TLM-FD algorithm based on the SCN is described in a series of publications [14–17]. A form of steady-state TLM based on a sinusoidal excitation and aimed at the automatic extraction of s-parameters in microwave circuits is described in Ref. [18].

In TLM-FD, a time harmonic variation is assumed, and thus only space discretization is necessary. Voltage waves incident on each port scatter and couple with other ports in a manner similar to that described for time-domain TLM (TLM-TD). It is easy to show that the scattering matrix is the same irrespective of the domain of formulation of TLM [17]. The aim is to formulate a set of equations relating the incident voltages on all ports to source (excitation) voltages. The process adopted to achieve this may be illustrated by reference to Fig. 10.7, where two SCN nodes like the one shown in Fig. 6.3 are shown schematically. The reflected voltage $V_1^r(x, y, z)$ at the center of node (x, y, z) is related to the incident

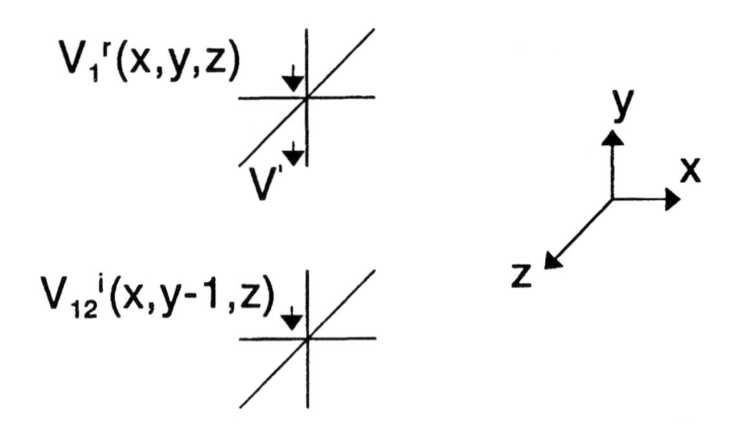

Fig. 10.7 Scattering and connection in TLM-FD

voltages at the center of the same node by the well known scattering matrix; i.e.,

$$V_1^r(x, y, z) = 0.5\left(V_2^i + V_3^i + V_4^i - V_{11}^i\right)$$

This pulse at its exit from the node (marked V′ in Fig. 10.7) is

$$V' = V_1^r(x, y, z)\, e^{-\gamma_1(x, y, z)\,\Delta y/2}$$

where $\gamma_1(x, y, z)$ is the propagation constant on line 1 of node (x, y, z). Similarly, the voltage incident on line 12 of node $(x, y-1, z)$ calculated at its center is

$$V_{12}^i(x, y-1, z) = V' e^{-\gamma_{12}(x, y-1, z)\,\Delta y/2}$$

In the above equation, $\gamma_{12}(x, y-1, z)$ is the propagation constant on line 12 of node $(x, y-1, z)$. It has been assumed that lines 1 of node (x, y, z) and 12 of node $(x, y-1, z)$ have the same characteristic impedance. This process is repeated until all nodes and ports have been covered to formulate the following matrix equation

$$\mathbf{A}\,\mathbf{V}^i = \mathbf{V}^s \qquad (10.10)$$

where $\mathbf{V}^s$ is a column vector defining the magnitude and phase of the source voltages, and $\mathbf{V}^i$ is the column vector of incident voltages on all ports in the problem. Matrix $\mathbf{A}$ is sparse of dimensions $12n \times 12n$, where n is the total number of nodes in the problem. Equation (10.10) may be solved for $\mathbf{V}^i$ using standard techniques such as the Jacobi or Conjugate Gradient methods. Excitation is applied in a similar manner as for TLM-

TD. As an example if component $E_x = E_0$ is to be excited on node (x, y, z), the following incident voltages are impressed on this node

$$V_1^i = V_2^i = V_9^i = V_{12}^i = \Delta x\, E_0/2$$

Output in the form of field components may be obtained from the TLM-FD model using formulae derived for TLM-TD; e.g.,

$$E_x = -\frac{\left(V_1^i + V_2^i + V_9^i + V_{12}^i\right)}{2\Delta x}$$

$$H_x = \frac{\left(V_4^i + V_7^i - V_5^i - V_8^i\right)}{2Z_0\Delta x} \tag{10.11}$$

where Z_0 is the impedance of the medium. The propagation constants on each line, which are required in order to formulate matrix A, can be calculated from the parameters of the medium and the TLM mesh. Using the subscript d to indicate quantities per unit length and subsequent subscripts to indicate line direction and polarization, the propagation constant on lines 1 and 12 is

$$\gamma_{yx} = \sqrt{j\omega L_{dyx}\,(G_{dyx} + j\omega C_{dyx})} \tag{10.12}$$

where L, G, and C represent inductance, conductance, and capacitance for these lines. Similarly, the characteristic impedance of these lines is

$$Z_{yx} = \sqrt{\frac{j\omega L_{dyx}}{G_{dyx} + j\omega C_{dyx}}} \tag{10.13}$$

The impedance of the medium is

$$Z_0 = \sqrt{\frac{j\omega\mu}{\sigma + j\omega\varepsilon}} \tag{10.14}$$

Imposing the condition that the characteristic impedance of all the lines is the same and equal to the impedance of the medium, and combining Equation (10.13) with (10.12) gives

$$(G_{dxy} + j\omega C_{dyx})\,\Delta y = \sqrt{\frac{\sigma + j\omega\varepsilon}{j\omega\mu}}\;\gamma_{yx}\Delta y \tag{10.15}$$

A similar expression is obtained for lines 2 and 9; i.e.,

$$(G_{dzx} + j\omega C_{dzx})\,\Delta z = \sqrt{\frac{\sigma + j\omega\varepsilon}{j\omega\mu}}\,\gamma_{zx}\Delta z \tag{10.16}$$

The total shunt admittance of all x-directed lines (1, 2, 9, and 12) is obtained by adding the right-hand side of the last two equations. Demanding that this sum is equal to the medium shunt admittance $(\sigma + j\omega\varepsilon)\Delta y\,\Delta z/\Delta x$, after some manipulation gives

$$\gamma S_x = \gamma_{yx}\Delta y + \gamma_{zx}\Delta z \tag{10.17}$$

where γ is the medium propagation constant, and $S_x = \Delta y\,\Delta z/\Delta x$. Expressions similar to Equation (10.17) are obtained for y- and z-directed lines. Demanding that lines on the same limb of the TLM node have the same propagation constant (i.e., $\gamma_{xy} = \gamma_{xz}, \gamma_{yx} = \gamma_{yz}, \gamma_{zx} = \gamma_{zy}$) and combining Equation (10.17) with the equivalent expressions for the y and z directions, after some manipulation gives

$$\gamma_{xy} = \gamma_{xz} = \frac{\gamma}{2\Delta x}(S_y + S_z - S_x)$$

$$\gamma_{yx} = \gamma_{yz} = \frac{\gamma}{2\Delta y}(S_x + S_z - S_y)$$

$$\gamma_{zy} = \gamma_{zx} = \frac{\gamma}{2\Delta z}(S_x + S_y - S_z) \tag{10.18}$$

These expressions provide all the necessary data for formulating A in Equation (10.10). It is clear from these expressions that the electrical properties of the medium may be varied by adjusting the propagation constants of each line without the need to introduce stubs. Stubs are required in TLM-TD to maintain synchronism, but in the frequency domain formulation described here, this complication has been removed. Further details of the method and its applications may be found in the references given, along with information regarding the relative merits of time- and frequency-domain TLM formulations for different problems.

10.6 IMPLEMENTATION ISSUES IN TLM

Throughout this book, emphasis was placed on developing the basic physical ideas associated with TLM modeling to a point where a solution algorithm may be developed to address a range of problems in electromagnetics and other areas. Most of the coverage was therefore concerned with model development and the TLM solver. However, in implementing and using TLM-based software, a number of other areas must be addressed, such as problem definition, preprocessing, output, and postprocessing of data. A time-domain differential code generates large amounts of data, so flexible, efficient postprocessing techniques are required to make sense of the information provided and present it to the user in an attractive manner. It is not possible to cover this interesting area to any depth in the present book, as it would require a major effort and anything said would, to some extent, be dependent on the hardware and software environment used for computation. It is, however, appropriate to address some general implementation issues that arise whenever TLM models are used.

Signal Processing

The TLM-TD model provides an output in the time domain consisting of the entire transient from the moment of the application of the excitation up to the termination of the computation. This information (e.g., field components) is available at all nodes in the model, but only a fraction of it is saved for further processing by the user. Time-domain data are frequently obtained from an impulsive excitation (duration Δt) applied to suitable locations in the model. Hence, the source is of a very broad frequency spectrum, exceeding the maximum frequency for which the model is valid. This upper frequency is normally taken to correspond to a wavelength equal to approximately 10 $\Delta \ell$. Other source excitation functions may be used, such as sinusoidal or a Gaussian pulse. In the latter case, there is more control of the frequency range in which energy is injected into the mesh. The Gaussian pulse is used extensively in FD-TD models to avoid instabilities at high frequencies. It is also useful in TLM models, although instability issues do not arise, when more control of the frequency spectrum is required. The Gaussian pulse in discrete form is given by the expression

$$f[n] = \exp\left(-\frac{n^2}{2s^2}\right) \tag{10.19}$$

where n is the time-step number, and s the standard deviation in time-steps. It is common to offset the pulse by n_0 time-steps from the time origin and select a standard deviation s such that there is insignificant truncation at early times, and a negligible amount of energy is injected at wavelengths shorter than a few mesh lengths $\Delta \ell$ (e.g., $n_0 = 60$, $s = 10$). Whichever method of excitation is used, a truncation of the time-domain signal is unavoidable. The larger the total number of time-steps in the simulation, the better the frequency resolution of the model will be. In high-Q systems, there is significant time-domain signal even after a very large number of time-steps. In such cases, it is profitable to adopt practices used in signal processing and apply a windowing function to the time-domain output prior to further processing to obtain frequency-domain data. For the type of data obtained from TLM simulations described in this book, a simple, one-sided linear window, where the time output is linearly reduced to zero over the last few hundred time-steps, was found to be adequate. Frequency-domain data can then be obtained using the Discrete Fourier Transform (DFT).

$$X(f) = \Delta t \left(\sum_{n=0}^{n-1} x[n] \right) \exp(-j\, 2\pi f\, n\, \Delta t) \tag{10.20}$$

The time-domain data $x[n]$ may be abruptly truncated at the end of the simulation (rectangular window), or in a more gradual way using the one-sided linear window described earlier. Whichever approach is used, the frequency response is effectively the convolution of the frequency spectrum of the signal with the frequency response of the window. For the case of a rectangular window, the frequency response is of the form $\sin(x)/x$, and this function is visible in many spectra obtained from simulations. As an example, the time-domain output for E_x inside a 1 m^3 air-filled cavity with conducting walls, subject to an impulsive excitation, is shown in Fig. 10.8a. For this simulation $\Delta \ell = 0.1$ m, and the total simulation time is 1,000 time-steps. The frequency domain data are shown in Fig. 10.8b. Using a one-sided linear window for the last 100 time-steps before applying the DFT gives the frequency spectrum shown in Fig. 10.8c. Comparing the last two figures shows that windowing has reduced the $\sin(x)/x$ terms and the dc offset in the frequency response. Running the simulation for a shorter time (600 time-steps) degrades the frequency response, as shown in Fig. 10.8d. The frequency step Δf used in the DFT must be chosen small enough that the maximum of the $\sin(x)/x$ curve can be adequately approximated. It is recommended that $\Delta f \leq 1.4/(\pi N \Delta t)$, where N is the total number of time-steps [19]. Attention must also be paid to the resolution of closely spaced resonances. Interference between the $\sin(x)/x$ terms associated with two adjacent resonance peaks can cause an error in the value of the resonance frequency. The argument of the $\sin(x)/x$ function is $\pi f N \Delta \mathrm{t}$. Hence, the maximum slope due to resonance 1 at the location of resonance 2 is $1/\pi F N \Delta t$, where F is the frequency difference between the two resonances. Near resonance 2, the slope is obtained by recognizing that

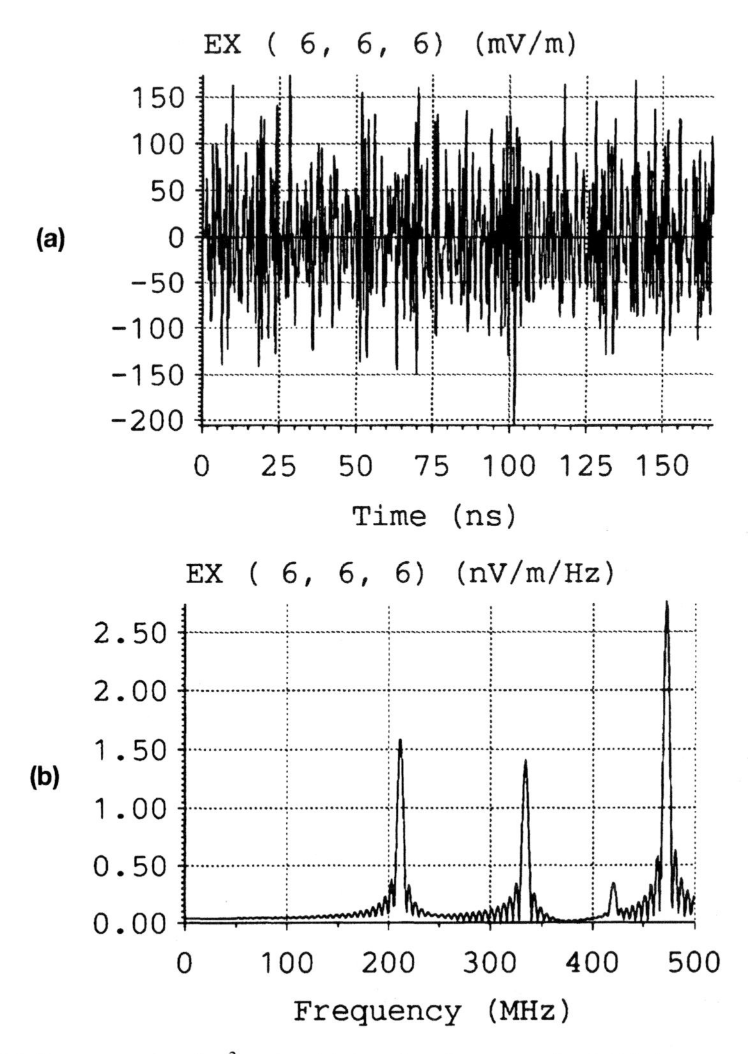

Fig. 10.8 Resonances in a 1 m^3 cavity: (a) time-domain output after 1,000 time-steps and (b) corresponding frequency spectrum *(continues)*

$$\frac{\sin x}{x} \cong \frac{x - x^3/6}{x} = 1 - x^2/6$$

Hence, the slope is

$$\left| \frac{d}{dx}(1 - x^2/6) \right| \cong \frac{x}{3} = \frac{\pi \Delta F \; N \; \Delta t}{3}$$

where ΔF is the deviation in frequency from the true resonance 2. Finding the stationary point on the combined response curve by equating the two slopes gives the frequency deviation of resonance 2 [20].

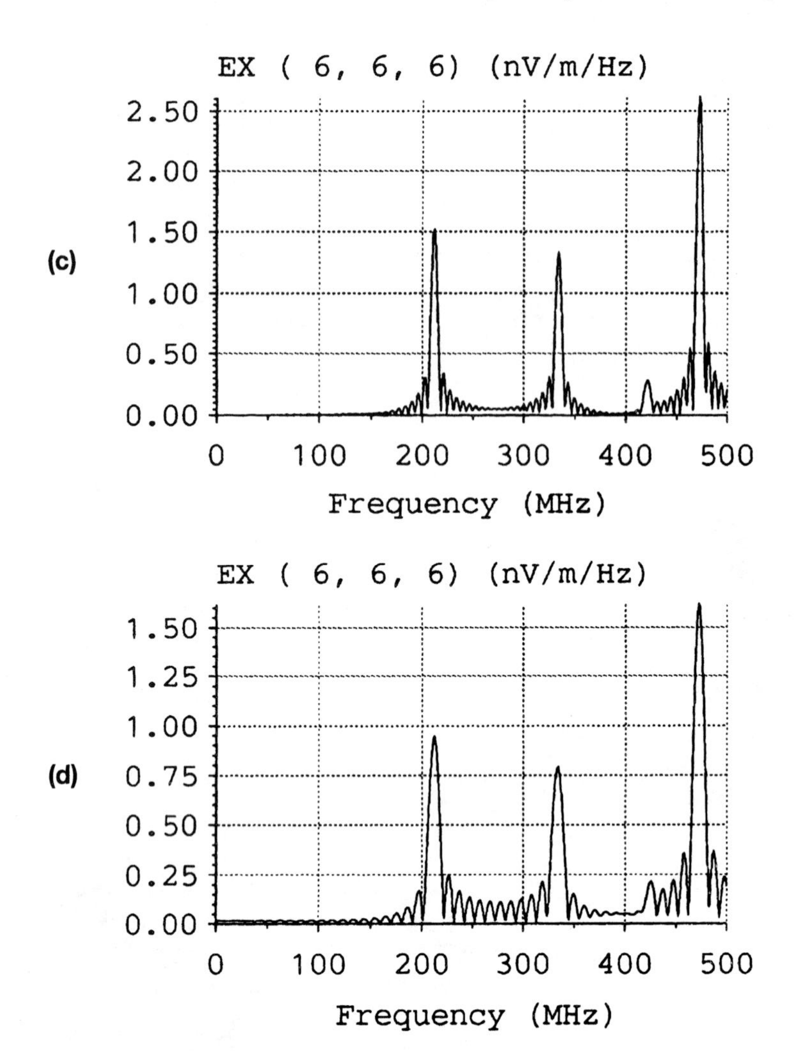

Fig. 10.9 *(continued)* (c) as in (b) but after windowing of the time-domain data, and (d) frequency spectrum obtained after 600 time-steps without windowing

$$\Delta F \cong \frac{3}{F\pi^2 N^2 (\Delta t)^2} \tag{10.21}$$

Further details on the spectral analysis of TLM data using techniques such as the Prony method may be found in Ref. [21].

Standard signal processing methods may be employed to time-domain data to remove high frequencies by using various forms of low-pass filtering. Once the impulse response $x[n]$ is known, the response $y[n]$ to any other time-domain excitation $f[n]$ may be obtained using discrete convolution.

$$y[n] = \sum_{i=0}^{n-1} x[i] f[n-i] \tag{10.22}$$

Computational Efficiency

The need to tackle increasingly larger and more complex problems has focused attention to increasing the efficiency of TLM algorithms by minimizing storage and run-time requirements. Although these matters will not be of immediate concern to someone first starting to use TLM, and any conclusions reached are likely to be dependent on hardware and software configurations, it is nevertheless useful to make some estimate of computational requirements. There is a trade-off between storage and run-time requirements in many situations. For example, if a graded mesh is used, a decision has to be made whether scattering parameters are precalculated and stored or calculated at each time-step. In many problems, it is possible to identify two or three regions where either the TLM meshing is different or the material properties vary. A TLM code that allows structuring of the problem into two or three blocks which are treated in the most efficient way can result in significant storage savings and improvements in execution speed. For problems that require large storage, advantage may be taken of virtual memory where a small physical address space is enhanced by swapping pages to and from a hard disk (paging). Rather than structuring the computation so that paging is used to complete a single scattering event for the entire problem, followed by connection and so on, it is more efficient to divide the mesh into two overlapping regions and complete a number of scattering and connections events within easy region before paging as explained in Refs. [19] and [22].

The efficiency of the main TLM scattering algorithm has steadily improved over the years. As an illustration, for the standard 12-port SCN, 36 additions/subtractions and 12 multiplications by 0.5 are normally required. If the calculation procedure described in Refs. [22] and [23] is used, then the number of multiplications is reduced to 6, but the total number of operations remains the same at 48. The modified scattering calculation procedure described in Ref. [24] brings down the number of additions/subtractions to 24 and multiplications to 6, giving a total number of operations equal to 30. For a general SCN capable of representing nonuniformities and electric/magnetic losses, 54 additions/subtractions and 12 multiplications are required. Depending on the nature of the mesh (regular SCN, stubbed SCN, HSCN, multigrid, or various combinations), average calculation times per node and time-step ranging from five to ten microseconds were recorded in a general-purpose workstation (49 SPEC-marks, 12 MFLOPS). Undoubtedly, careful structuring of the software and calculation procedures and exploitation of parallel computer architectures will result in faster calculation times and enhanced simulation capabilities.

Dispersion and Choice of Discretization Length

Dispersion in a 2D TLM mesh was discussed in some detail in Section 5.3. Similar principles apply for dispersion in a 3D TLM mesh, but the extend of dispersion at high frequencies depends on the type of node used. Closed form algebraic expressions have been obtained for some nodal configurations, but so far most studies of dispersion have been numerical in nature. Details may be found in Refs. [25] and [26].

A better view of propagation in a 3D TLM mesh consisting of regular symmetrical condensed nodes may be obtained by plotting the wavenumber vector normalized to the medium wavenumber, as shown in Fig. 10.9a, for the ideal case of a very fine discretization ($\Delta\ell/\lambda_0 \to 0$) and for two finite values of discretization $\Delta\ell/\lambda_0 = 0.1$ and 0.2. The percentage error for different angles of propagation is shown more clearly in Fig. 10.9b.

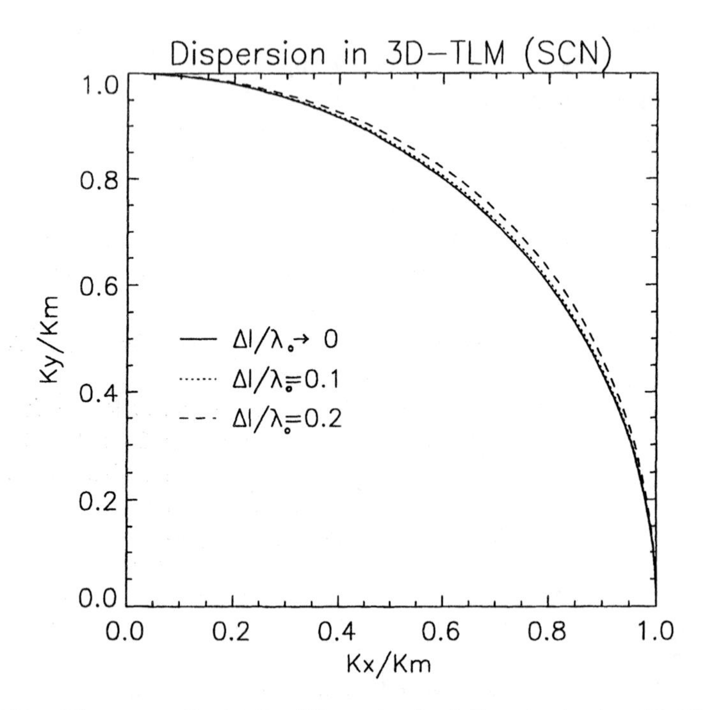

Fig. 10.10a Wave vector (k_x, k_y) for different levels of discretization in a 3D SCN TLM mesh

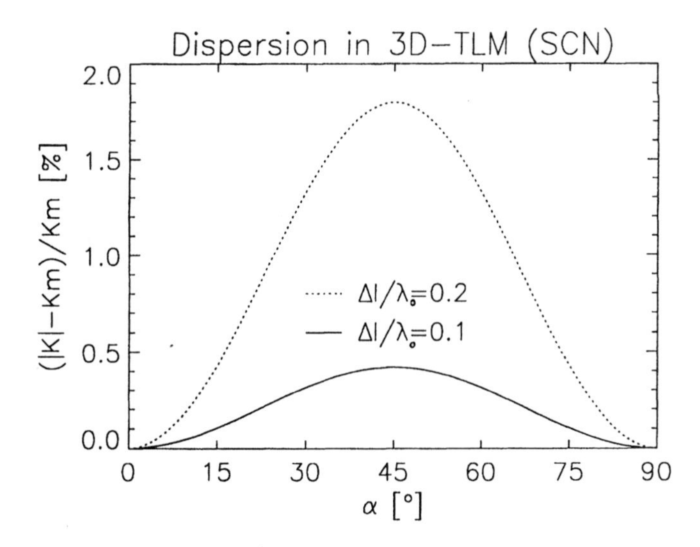

Fig. 10.11b Percentage error due to dispersion for different angles of propagation in a 3D SCN TLM mesh

From these graphs, it is clear that using a $\Delta\ell$ smaller than a tenth of the wavelength at the highest frequency of interest results in errors lower than 1 percent, which is an acceptable level in most applications.

REFERENCES

[1] Holland, R., and L. and Simpson. 1981. Finite-difference analysis of EMP coupling in thin struts and wires. *IEEE Transactions* EMC-23, 88–97.

[2] Naylor, P., and C. Christopoulos. 1990. A new wire node for modelling thin wires in electromagnetic field problems solved by transmission line modelling. *IEEE Transactions* MTT-38, 328–330.

[3] Wlodarczyk, A.J., and D.P. John. 1992. New wire interface for graded 3-D TLM. *Electronics Letters* 28, 728–729.

[4] Porti, J.A., J.A. Morente, M. Khalladi, and A. Callego. 1992. Comparison of thin-wire models for TLM method. *Electronics Letters* 28, 1910–1911.

[5] Morente, J.A., J.A. Porti, G. Gimenez, and A. Gallego. 1993. Loaded-wire node for TLM method. *Electronics Letters* 29, 182–184.

[6] Jordan, E.C., and K.G. Balmain. 1966. *Electromagnetic Waves and Radiating Systems,* 2nd ed. Englewood Cliffs, NJ: Prentice Hall, 513–519.

[7] Johns, D.P., A. Mallik, and A.J. Wlodarczyk. 1992. TLM enhancements for EMC studies. *Proceedings of the 1992 IEEE Regional Symposium on EMC* (Tel Aviv), 3.3.3/1–6.

[8] Johns, D.P., A.J. Wlodarczyk, and A. Mallik. 1991. New TLM models for thin structures. *Proceedings of the International Conference on Computation in Electromagnetics* (London), IEE Conf. Publ. 350, 335–338.

[9] Cangellaris, A.C., and D.B. Wright. 1991. Analysis of the numerical error caused by the stair-stepped approximation of a conducting boundary in FDTD simulations of electromagnetic phenomena. *IEEE Transactions* AP-39, 1518–1525.

[10] Johns, P.B., and G.F. Slater. 1973. Transient analysis of waveguides with curved boundaries. *Electronics Letters* 9, 486–487.

[11] Muller, U., and W.J.R. Hoefer. 1992. The implementation of smoothly moving boundaries in 2D and 3 D TLM simulations. *Proceedings of the IEEE Microwave Symposium* (Albuquerque, NM) 791–792.

[12] German, F.J. 1993. Infinitesimally adjustable boundaries in symmetrical condensed node TLM simulations. *Proceedings of ACES '93,* 9th Annual Review of Progress in Applied Computational Electromagnetics (Monterey, California) 482–490.

[13] Brewitt-Taylor, C.R., and P.B. Johns. 1980. On the construction and numerical solution of transmission-line and lumped network models of Maxwell's equations. *International Journal for Numerical Methods in Engineering* 15, 13–30.

[14] Johns, D.P., A.J. Wlodarczyk, A. Mallik, and C. Christopoulos. 1992. New TLM technique for steady-state field solutions in three-dimensions. *Electronics Letters 28,* 1692–1694.

[15] Johns, D.P., and C. Christopoulos. 1993. Lossy dielectric and thin lossy film models for 3D steady state TLM. *Electronics Letters* 29, 348–349.

[16] Johns, D.P., and C. Christopoulos. 1993. Dispersion characteristics of 3D frequency-domain TLM. *Electronics Letters* 29, 1536–1537.

[17] Johns, D.P., and C. Christopoulos. 1994. A new frequency-domain TLM method for the numerical solution of steady-state electromagnetic problems. *IEE Proceedings Sci. Meas. Technol.* 141, 310–316.

[18] Jin, H., and R. Vahldieck. 1992. The frequency-domain transmission-line matrix method—a new concept. *IEEE Transactions* MTT-40, 2207–2218.

[19] Herring, J.L. 1993. Developments in the Transmission-Line Modelling Method for Electromagnetic Compatibility Studies. Ph.D. thesis, University of Nottingham, England.

[20] Johns, P.B. 1972. Application of the transmission-line matrix method to homogeneous waveguides of arbitrary cross-section. *Proceedings of the IEE* 119, 1086–1091.

[21] Wills, J.D. 1990. Spectral estimation for the transmission line matrix method. *IEEE Transactions* MTT-38, 448–451.

[22] Herring, J.L., and C. Christopoulos. 1993. The application of different meshing techniques to EMC problems. *Proceedings of the 9th Annual Review of Progress in Applied Computational Electromagnetics* (Monterey, California), 755–762.

[23] Naylor, P., and R. Ait-Sadi. 1992. Simple method for determining 3D TLM nodal scattering in nonscalar problems. *Electronics Letters* 28, 2353-2354.

[24] Trenkic, V., C. Christopoulos, and T.M. Benson. 1994. New developments in the numerical simulation of RF and microwave circuits using the TLM

method. *Facta Universitatis (Nis),* Series: Electronics and Energetics 7, 61–66.

[25] Nielsen, J.S., and W.J.R. Hoeffer. 1991. A complete dispersion analysis of the condensed node TLM mesh. *IEEE Transactions on Magnetics* 27, 3982–3985.

[26] Krumpholz, M., and P. Russer. 1994. On the dispersion in TLM and FDTD. *IEEE Transactions* MTT-42, 1275–1279.

Index